AN INTRODUCTION TO
DEVELOPMENTAL BIOLOGY

AN INTRODUCTION TO DEVELOPMENTAL BIOLOGY

JOHN MCKENZIE M.D.

Reader in Developmental Biology
University of Aberdeen

BLACKWELL SCIENTIFIC PUBLICATIONS

OXFORD LONDON EDINBURGH MELBOURNE

© 1976 Blackwell Scientific Publications
Osney Mead, Oxford
85 Marylebone High Street, London, W1M 3DE
9 Forrest Road, Edinburgh
P.O. Box 9, North Balwyn, Victoria, Australia

ISBN 0 632 07820 0

First published 1976

Distributed in the U.S.A. by
Halsted Press, a Division of
John Wiley & Sons Inc
New York

Set by Hope Services Ltd and
Printed and bound in Great Britain by
Butler and Tanner, Ltd., Frome

CONTENTS

PREFACE

Introducing a subject to students for the first time is always something of a challenge but to stimulate and maintain their interest throughout that critical phase may be perhaps even more difficult; of course, everyone has his (or her) own special method or technique which he (or she) likes to think is above reproach. Suffice it to say, then, that this text is based on many years experience of either presenting developmental biology in a general and palatable form to supplement other studies in the field of biology or of introducing students to a more intensive study of the various aspects of the subject, leading to an Honours degree. Naturally, there were modifications of the programme with each passing year but the broad pattern has remained the same; breadth of study rather than depth was the aim, with detail varying according to the length of the course but seldom reaching the overall standard provided here. It should not be difficult for either the student or teacher to select according to whether he wishes a cordial acquaintance with the subject or a foundation for further study.

I have always had a genuine sympathy for those students on whom I inflicted my earlier attempts to provide an interesting course in developmental biology as well as for those of later years when modifications were being introduced; but I am for ever indebted to them; wittingly and unwittingly they moulded the course by their reactions, often critical but never unkind, and they take more credit than they realize. I am equally grateful to all my colleagues past and present for their help, tolerance and criticism; only I can judge how much and how well they have contributed.

I must however mention especially Dr Frank Cusick for his welcome comments and criticisms of the chapter on the development of plants; my wife, who suffered cheerfully the long tedious hours I spent reading and writing for the text; also Miss Jean R. Grant for her patient work in typing (and retyping) the manuscript, as well as for the other innumerable tasks involved in the preparation of text and illustrations; and, not least, Mr Robert Campbell of Blackwell Scientific Publications whose guidance, exhortations and patience were admirably applied and genuinely appreciated by the author. Yet, while I acknowledge so much reliance on those around me, I hold myself entirely responsible for any deficiencies you may find in the pages which follow.

1 INTRODUCTION

If you have an inquisitive frame of mind, then try studying developmental biology; your curiosity will surely be satisfied. All forms of life—animal and vegetable, large and small—undergo development but, because the changes are so gradual, we scarcely notice them, taking them for granted and failing to appreciate their mysterious ways; only when development is abnormal or when, for some reason, we are relying on it, does our attention become focused on the normal process.

Nowadays, an understanding of normal development is being increasingly sought; as well as the intrinsic value of being able to explain its mysteries, the biology of development has its applications in the many and varied walks of life such as food production, agriculture, fisheries and in different aspects of medical and veterinary work. For rearing cattle and growing crops along the most economical and profitable lines the farmer applies the principles of developmental biology—often empirically—but the better his awareness and the more efficient his application of these principles the more he and the community will benefit. Considerable advances have been made in that direction already and, in another context, namely fisheries, increasing attention is being paid to improving the harvest from seas and rivers. Failure of growth or development in crops or distortion of development in an animal is naturally a financial loss of varying magnitude. However, when a child fails to grow or is born with a congenital defect, the consequences are more serious and for the parents it is nothing short of a tragedy.

These are only a few examples of the importance of development in the community but in many other less obtrusive instances, development or mal-development leaves its mark in an indelible fashion. To be able to control development, to improve growth and development and to prevent failures and abnormalities are our main aims in learning and researching in developmental biology but there is a long way to go before these are achieved. Therein lies the challenge for the developmental biologist.

The fundamental principles of developmental biology are the same for every living creature and plant and can be exemplified in any of them. The complexities of development like the complexities of life are graded in animals and plants and theoretically we should begin by examining the development of the simplest organisms and progress to the higher. In some measure, this is desirable, perhaps essential, but on the assumption that most people are interested more in themselves and their immediate environment than in the obscure and less familiar examples of life, the aim of this introduction to developmental biology is to direct the students' attention as soon as possible to features in their own development and in animals and plants with which they are all familiar. Once their interest has been aroused by the relevance of developmental biology, further exploration and thereafter research with experimentation will continue wellnigh spontaneously on those forms of life which will more readily provide the answers to questions posed by the enquiring mind.

Developmental biology is only one aspect of biology and by no means a clearly-defined, distinct element of that wider study. Ideally the student should first have a broad background including, for instance, cell biology, histology, basic anatomy, physiology and biochemistry and genetics as well as general biological principles. Depending on the particular aspect(s) of development followed by him as a more senior student, he must acquire and apply other disciplines such as molecular biology, advanced biochemistry, pathology, pharmacology and special techniques like cell culture for further study and research.

Indeed, much of what has been learned in the recent past and what will probably form the guide lines in future research has involved and will continue to involve principles and techniques drawn from other disciplines. Yet this does not imply a subordinate role and heavy reliance by developmental biology on other established disciplines. On the contrary, it has a strong unifying theme with many ramifications into areas of biology like reproduction, endocrinology, behaviour, learning and nutrition as well as its applications in industry and the professions. No one doubts its importance in biology since it occupies a central position. Indeed the serious study of developmental biology itself provides an unusually valuable training and background in biology generally.

2 THE CELL

Under the microscope, a cell is no more than a tiny mass of inanimate material; it displays a wealth of intricate detail when examined with the electron microscope (EM) and it is possible to identify innumerable chemical substances located in specific regions; but to think of a cell in these terms only is a mistake. You are looking at a living structure which died while being prepared for histology and all that can be seen is a snapshot of one moment in the life of a cell—processed to bring up whatever details are being sought. The cell is, in fact, a unit of living matter in much the same way as human beings are, i.e. capable of living an entirely independent existence if forced into such a predicament. Cells also differ in their structure and function within their society just as people differ in their size, shape and contribution to human society. When we realize, however, that all the different cells comprising a multicellular organism have been derived from the single cell formed by the fusion of two sex cells and then ask how this is achieved, we have broached *the* fundamental problem in developmental biology—namely the mechanisms of *cell differentiation.*

The basic cell

The concept of a *basic* or *typical cell,* however useful, is probably only theoretical because it is very doubtful if such a cell even exists; it may be visualized, however, as having the basic features common to every cell but lacking the modifications of the specialized or differentiated cells. Cells differentiate for one reason only, namely to carry out specialized functions but these are not entirely new acquisitions because even the basic cell is capable in some measure, often very small, of behaving like differentiated cells. Differentiation is, in other words, the enhancement of one or more of a cell's basic faculties until it is an expert in that particular activity; and expertise depends on morphological, physiological and biochemical modifications—not fundamental alterations—of the basic cell.

Cell components and functions (Fig. 2.1)

Only those features of the basic cell relevant to development and differentiation will be mentioned here. Without its *nucleus* the cell soon dies; inside it are the *chromosomes,* long threads consisting mostly of deoxyribonucleic acid (DNA) but so thin that they are invisible with the light microscope except during

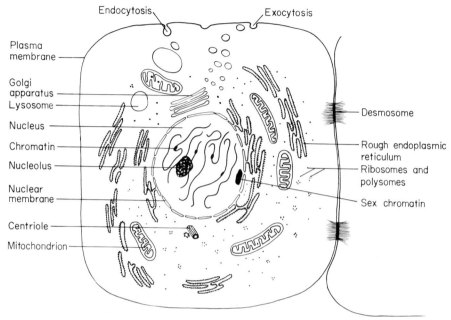

Fig. 2.1 The basic or typical cell (animal).

cell division. Scattered throughout the nucleus are flecks of *chromatin* which consist of tightly coiled segments, the inactive parts, of chromosomes. One or two *nucleoli*, large darkly-stained bodies, can also be seen, consisting of ribonucleic acid (RNA) and protein and concerned with protein synthesis. The *sex chromatin*, a small easily stained particle, lying close to the nuclear membrane within the nuclei of cells from female animals, is the inactive member of the two X chromosomes. The rest of the nucleus is occupied by *nuclear sap*, a colloidal solution containing the usual organic substances and bathing the other constituents.

The real nature and extent of the *membrane system* of a cell has been understood only since the introduction of the EM. The whole cell is enclosed in a *unit* or *plasma membrane*, 75Å thick but consisting, like other parts of the system, of three layers, the middle one mostly of lipid and the other two mostly of protein. Water and electrolytes diffuse easily through the plasma membrane but, for some substances, passage through it probably involves their chemical combination with its lipoprotein. Larger molecules enter the cell by pinocytosis (see below). The *nuclear membrane* consists of *two* unit membranes separated by a very narrow space but fused at innumerable points to produce *nuclear pores*. Most people believe that these are actual channels between nucleus and cytoplasm allowing the easy transfer of molecules. A regular feature of the cytoplasm is the *endoplasmic reticulum* (ER), an extensive system of flattened intercommunicating vesicles. Usually their outer surfaces are studded with RNA-containing granules, the *ribosomes*; these *rough-surfaced vesicles* take part in the synthesis of enzymes and other secretions and, when present in large numbers and stained with basophilic dyes, they form the *chromidial substance*. The newly manufactured material, located first in those vesicles, is later transferred to the *Golgi apparatus* lying near the nucleus. This tiny complex comprises three kinds of vesicles, in continuity with one another and with the rough-surfaced vesicles. Secretions are concentrated there and perhaps elaborated further e.g. conjugated with carbohydrate. Many of the enzymes manufactured by the cell would readily attack the other constituents of the cell but, apparently, the unit membranes of the endoplasmic reticulum and of the Golgi apparatus can isolate the enzymic material and protect the rest of the cell. Some vesicles of the Golgi complex migrate, with their secretion, towards the surface of the cell where the wall of each vesicle fuses with the plasma membrane; the fused portions then break down to release the secretion from the cell. In this way, the continuity of the unit membrane enclosing the cell is maintained. The process is often referred to as reversed pinocytosis or *exocytosis*; normal *pinocytosis* or *endocytosis*, the method whereby large molecules enter the cell, begins by the material sinking into a depression of the plasma membrane; as the pocket in the membrane deepens to enclose the material, its edges close, shutting off the vesicle from the surface and restoring the continuity of the surface membrane.

When dead, dying or foreign particles have to be disposed of, some cells are ready to engulf them by *phagocytosis*, a process somewhat similar to pinocytosis. Whether cells are forewarned and forearmed to digest such material is not clear but many have vesicles called *lysosomes* containing hydrolytic enzymes. These vesicles meet and fuse with the vesicle containing the phagocytosed material which is then digested and destroyed.

Free ribosomes, found scattered throughout the cytoplasm, are involved in the manufacture of *structural proteins* like the contractile muscle proteins and haemoglobin, which, of course, remain in the cell. Some parts of the ER, the *smooth-surfaced vesicles*, are devoid of ribosomes but their function is not clear.

Mitochondria, (sing. mitochondrion) tiny, round or oval bodies scattered throughout the cytoplasm also require electron microscopy to display them satisfactorily. Each has two unit membranes, an outer one completely enclosing an inner and separated from it by a narrow space; the inner membrane is folded to produce an irregular series of shelves, the *cristae mitochondriales*, projecting into the cavity of the mitochondrion. A few dark-staining granules containing DNA are often seen inside the mitochondria. Biochemical analysis of mitochondria reveals that they contain the substrates and enzymes of the citric acid (Krebs) cycle and that cell respiration occurs there. In other words, the aerobic phase of carbohydrate breakdown from pyruvic acid onwards via the Krebs cycle occurs in the mitochondria. The energy obtained in this way is used by another group of enzymes in the mitochondria to add an inorganic phosphate group to adenosine diphosphate (ADP) and thus form adenosine triphosphate (ATP) as in the process of *oxidative phosphorylation*. The energy is stored in the form of ATP, carried to wherever it is required in the cell and released by adenosine

triphosphatase which splits off the phosphate group acquired in the mitochondrion. There may still be some doubt whether all the enzymes involved in these
reactions are present in the mitochondria. Even more obscure are, first, the site
of manufacture of these enzymes and second, the mechanics of transporting the
ADP, ATP and the necessary substrates into and out of the mitochondria.
Nevertheless here is a nest of enzymes segregated from the rest of the cytoplasm
by a unit membrane and producing most or all of the energy required by a cell.

Close to the nucleus, in a little pale area called the *centrosphere* or *centrosome*,
the EM reveals two tiny structures, the *centrioles*; each is a cylinder open at
both ends and having as its wall nine sets of three very fine tubules. Centrioles
are concerned with cell division and related to the development of cilia.

All the structures described so far are bathed in an apparently homogeneous
solution consisting mostly of water containing the usual salts, electrolytes, small
protein molecules, polypeptides, amino acids, carbohydrates and lipids necessary
for normal metabolism. But the cytoplasm often contains visible quantities of
these substances. Carbohydrate, for instance, is transformed into the large
polysaccharide molecule, glycogen, and stored in this way in liver cells and in
muscle. *Protein* does not accumulate like glycogen but the large protein
configurations of differentiated cells will be discussed later. Fat is frequently
stored and appears first in the form of tiny droplets distributed through the
cell; as it increases in amount, the droplets coalesce and with further increase,
it compresses the cytoplasm and then the nucleus until both are flattened
against the plasma membrane.

THE PLANT CELL (Fig. 2.2)

A typical plant cell has all the features of the animal cell except centrioles but it
also has a few specialities of its own. *Plastids*, found only in plants, are of three

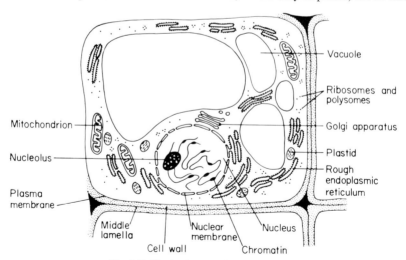

Fig. 2.2 The basic or typical cell (plant).

types: *chloroplasts* are disc-shaped granules, which consist of membranous lamellae studded with chlorophyll molecules giving a green colour; they are responsible for photosynthesis i.e. the manufacture of food under the stimulus of light and the incorporation of radiant energy into sugar. In addition to chlorophyll, there are also small amounts of the pigments, carotene and xanthophyll, but in chromoplasts there is enough of these yellow-orange or red pigments to give the colours found in the carrot and in the tomato. The third type, the *leucoplast*, has no pigment, contains starch grains and is usually found in cells of roots and tubers. The *vacuoles* in the cytoplasm are primarily dilatations in the ER but they are often so large that they distort the internal architecture of the cell pushing nucleus and cytoplasm close to the periphery. Other inclusions in plant cells reflect their synthetic activities e.g. *starch grains*, *silica*, *calcium salt crystals* and *oil* or *fat* droplets.

Protein synthesis (Fig. 2.3)

The process of cellular protein synthesis can be traced back to the code provided by the sequence of bases, thymine, adenine, guanine and cytosine in

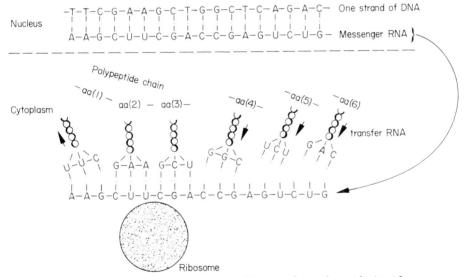

Fig. 2.3 Scheme illustrating the main features of protein synthesis. After Ham, A. W., *Histology*, 5th edition. Pitman Medical Publishing Co. Ltd., London.

the nucleotides comprising the strands of the DNA molecule. The code is read in groups of three (triplets) along one of the strands and is transcribed to the strand of messenger ribonucleic acid (mRMA) which forms against the DNA. Ribonucleic acid possesses the same bases as deoxyribonucleic acid except for the substitution of uracil for thymine and when it disengages from the DNA, the mRNA moves to the ribosome where the triplet code is translated in the following way. Amino acids in the cytoplasm are picked up by special molecules, soluble or transfer RNA (tRNA) and there is one molecule of tRNA specific for each amino acid. Each of these tRNA molecules also corresponds to a specific triplet of bases on the mRNA strand and, as these tRNA molecules become arranged according to the code, so the amino acids attached to them join up with one another to form polypeptides, the first step towards the formation of a protein. Where the ribosomes are adhering to the endoplasmic reticulum the newly formed protein is found in the ER vesicles and eventually secreted by or excreted from the cell. Where the ribosomes are free in the cytoplasm the protein remains in the cell.

MITOSIS (Fig. 2.4)

No cell can live indefinitely; it either dies (eventually) or divides and, for the survival of the speicies, be it amoeba or human, some sort of division of

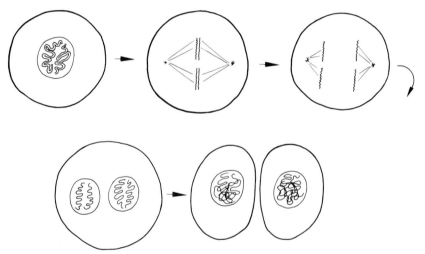

Fig. 2.4 Series of diagrams illustrating the main features of mitosis; only two chromosomes are shown.

nucleus and cytoplasm is essential. Apart from this general principle, however, the need for cell division, particularly mitosis, is greater during growth and development and is usually thought of in this context.

Prior to mitosis, i.e. in the *interphase* cell, the two strands of nucleotides forming each chromosome separate to allow the formation of a new strand of nucleotides alongside each parent strand. In this way, each chromosome is replicated and each double strand in the replicated chromosome is known as a *chromatid.* During the different stages of *mitosis—prophase, metaphase, anaphase* and *telophase,* (terms introduced to describe the gradually changing features and not as an indication of abrupt transitions), the cell divides into two equal and identical daughter cells in a smoothly flowing process. The chromosomes coil and shorten until they are visible by the light microscope, the nucleoli fade and the nuclear membrane disappears. The two newly formed pairs of centrioles (each centriole replicates during interphase) migrate to opposite sides of the chromosome cluster; fine microtubules radiate from the new centrospheres, many of them obtaining an attachment to the chromosomes as they line up in a metaphase (equatorial) plate, the whole assembly forming a *mitotic spindle*. Separation of the paired chromatids follows—perhaps as a result of contraction of the microtubules—and each chromatid becomes a chromosome. Thus two complete and equal sets of chromosomes migrate in opposite directions towards the centrosphere. A new nuclear membrane forms around each group and nucleoli reappear. Each chromosome uncoils and becomes invisible, whilst division of the cytoplasm to create two separate cells occurs by the formation of a constriction furrow in the plasma membrane and the development of new membranes between the two nuclei. Since each chromosome has been duplicated exactly and in its entirety, each daughter cell is an exact genetic replica of its parent. In cell division, the accompanying replication of the chromosomes and the division of nucleus are not the only features. There are others equally intriguing but for the most part poorly understood viz. the increase in the surface area of the plasma and intracellular membranes, the increase in the number of mitochondria and plastids and of centrioles necessary for the two daughter cells.

The increase in cell membrane probably occurs by the intercalation of new molecules into the pre-existing membrane but whether the process is confined to certain localized areas or can occur throughout the membrane is uncertain. At the time of cell division fresh membrane appears to be formed in the cleavage furrow but it probably also increases in extent elsewhere during the interphase period at a steadier pace.

Mitochondria are believed to increase in number by growth in size and division of existing mitochondria while plastids probably multiply in the same way in dividing plant cells while they are still at the proplastid stage.

Each new centriole seems to develop near one end of a mature centriole and grow at right angles to it. When the two mature centrioles separate in mitosis, each takes its own newly-formed centriole with it.

The stimulus and requisite conditions for cell division have also been the subject of investigation but so far no clear explanation has emerged.

MEIOSIS (Fig. 2.5)

Most organisms reproduce by the formation of sex cells i.e. ova and spermatozoa, which fuse to form the new generation. If these sex cells (*gametes*) were produced by mitosis, each would have the same number of chromosomes as its parent;

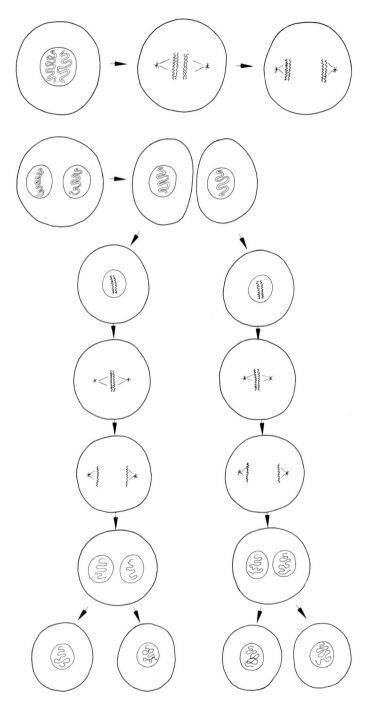

Fig. 2.5 Meiosis: a series of diagrams illustrating the main features of the two cell divisions: only two chromosomes are depicted in the parent cell.

then, with fusion of gametes at fertilization, the newly formed *zygote* would have twice as many chromosomes as the parents and, without some method of reducing the number by half at some stage in the life cycle, the situation with regard to chromosome numbers would soon become ridiculous. Reduction division or meiosis solves this problem and also provides the mechanism of sex determination and a method of shuffling and redistributing genetic material. Chromosomes are paired (e.g. 46, or 23 pairs in the human) and the full complement is the *diploid* number (2n). Meiosis, which reduces this to the haploid count (n) by distributing the members of each pair of chromosomes to one or other of the two daughter cells, usually occurs during development of the gametes, so that these are the only haploid cells in the life cycle. Exceptions occur, as in some lower plants e.g. algae, where individual haploid cells (meiospores) multiply to produce a stage (the gametophyte) in the life cycle of the plant in which all the cells are haploid. Release of gametes from the gametophyte is followed by fusion of gametes from different plants and thereafter the development of the sporophyte which is similar in appearance to the gametophyte: the cells released from the sporophyte undergo meiosis first and become the haploid meiospores.

Meiosis, in fact, always consists of two successive cell divisions, in the first of which the actual reduction in the chromosome number occurs; the second is fundamentally similar to mitotic division. The same stages can be identified in both divisions but prophase 1 (i.e. in the 1st division) is prolonged and has stages of its own. At the end of prophase 1/beginning of metaphase 1, the chromosomes, each as a pair of chromatids, arrange themselves in their *homologous pairs* on the metaphase plate. There are therefore four chromatids apposed to one another along their lengths. At the end of metaphase, separation of the chromosome pairs occurs—one chromosome of each pair to each new nucleus—the chromatids of each chromosome remaining together. In this way, each new daughter cell receives only one from each chromosome pair i.e. the haploid number. During the second division, each chromosome divides as in mitosis and the final result of the two divisions is four haploid daughter cells.

Each cell in the female organism has two identical sex chromosomes (XX) and therefore each ovum will have one X chromosome. In the male however, the two sex chromosomes are not identical—one is X (similar to those of the female) but the other, known as Y, is smaller and easily distinguished from its partner. Each spermatozoon receives one or other of these chromosomes and therefore half of them have an X and half of them a Y chromosome. When fertilized, the ovum thus acquires either an X or a Y chromosome and the zygote will have either the XX(female) or XY(male) set. The condition in which the female gamete contains two similar sex chromosomes and the male dissimilar occurs in most animals but in some cases, e.g. birds, the reverse is true.

At fertilization, the organism receives, for every chromosomal pair, one from the father and one from the mother but, when its own germ cells are undergoing the first stage of meiosis, the 'paternal' chromosomes need not, and do not, all proceed to the same daughter cell and the 'maternal' to the other;

the chromosomes are 'shuffled' i.e. irregularly orientated so that 'maternal' and 'paternal' chromosomes are distributed at random between the two new nuclei at the end of metaphase. Furthermore, during prophase 1, when the chromatids of the two homologous chromosomes are all close to each other, several points of contact occur and the chromatids may become 'entangled', i.e. they may cross one another at several points. But the chromatids also break at these points and unite again in such a way that they effectively exchange segments i.e. genes, with each other. Thus the chromatids may not emerge from meiosis with the same genetic constitution as they entered. During meiosis, therefore, in addition to a shuffling of the cards (chromosomes), there is also a shuffling of the spots (genes) and random genetic variation for the next generation is assured.

FURTHER READING

Ambrose E.J. & Easty D.M. (1970) *Cell Biology*. London: Thomas Nelson & Sons Ltd.

Berrill N.J. (1971) *Developmental Biology*. New York & London: McGraw-Hill Book Company.

Dyson R.D. (1974) *Cell Biology—A Molecular Approach*. Boston: Allyn and Bacon, Inc.

Ham A.W. (1965) *Histology* 5th Edition, Chapters 5, 6 & 7. London: Pitman Medical Publishing Co. Ltd.

Ingram V.M. (1964) *Biosynthesis of Macromolecules* 2nd ed. Menlo Park, California: W:A. Benjamin, Inc.

Nason A. & De Haan R.L. (1973) *The Biological World* Part 3. New York & London: John Wiley & Sons Inc.

Porter K.R. & Bonneville M.A. (1968) *Fine Structure of Cells and Tissues* 3rd ed. Philadelphia: Lea & Febiger.

Spratt N.T. Jr. (1971) *Developmental Biology*. Belmont, California: Wadsworth Publishing Company, Inc.

The Open University (1974) *Physiology of Cells and Organisms* Units 1, 3, 4 & 7. Milton Keynes: The Open University Press.

Torrey J.F. (1967) *Development in Flowering Plants*. London: The Macmillan Company, Collier-Macmillan, Ltd.

3 INTERCELLULAR SUBSTANCES

Between the cells is a host of different materials–all of them important to the cell and to the tissues they form. *Tissue fluid*, in different concentrations and composition according to the tissue, permeates the whole body and contains all the normal physiological constituents. The other intercellular substances, best considered in two groups, the *formed* or *fibrillar* and the *amorphous*, are products of the cells associated with them.

FIBRILLAR INTERCELLULAR SUBSTANCES

There are three kinds of this material–*collagen, reticular*, and *elastic fibres*. Fresh *collagen fibres* (Fig. 3.1) are white but, when stained with haematoxylin and eosin (H & E), appear as pink strands of different diameters, and homogeneous apart from a few faint longitudinal striations. Each collagen fibre can be teased into thinner collagen fibrils and, with the EM, the fibrils are seen to consist of even finer threads, the microfibrils; at this magnification the whole length of the microfibril shows alternate light and dark bands (periodicity), due to the arrangement of the constituent tropocollagen molecules in them. In more delicate, more cellular, tissues where strong support is unnecessary, the framework consists of fine *reticular fibres*. Although they do not stain with H & E and give, instead, a positive reaction with Periodic acid-Schiff reagent (PAS), these fibres are probably of similar chemical composition to collagen since the EM has revealed that they have the same periodicity; the positive PAS reaction is probably due to their conjugation with intercellular glyco-protein. There is no doubt that *elastic fibres* are chemically different; they are homogeneous, capable of recoil when stretched and give the tissues a yellow colour. Special stains, e.g. orcein, are needed to demonstrate them but even without staining they can be seen as glassy refractile threads. In some parts of the body, elastic tissue occurs as multilaminated sheets or membranes.

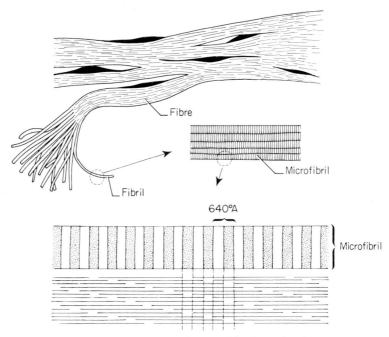

Fig. 3.1 Collagen at successively higher magnifications; also the arrangement of the tropocollagen molecules responsible for the periodicity of the micro-fibril. After Ham, A. W. *Histology* 5th edition. Pitman Medical Publishing Co. Ltd., London.

AMORPHOUS INTERCELLULAR SUBSTANCES

These form a family of *protein-carbohydrate complexes*, only a few of which have been fully investigated or even clearly identified. *Hyaluronic acid*, for instance, an acid mucopolysaccharide containing glucosamine and glucuronic acid, is a soft jelly-like material with the ability to 'hold' water. In addition to forming an embedding medium for collagen, it also prevents the invasion and spread of bacteria and dyes through the tissues. *Chondroitin sulphate* forms a firm gel and gives a tougher, even rubbery, consistence to a tissue e.g. cartilage. Both can be identified by their metachromatic reaction to dyes like toluidine blue and methylene blue. The *glycoproteins*, which contain large quantities of hexosamine, are usually associated with reticular fibres and basement membranes giving a positive reaction with the PAS technique.

PLANT INTERCELLULAR MATERIAL

Intercellular substances are just as important in *plants*; they are represented by the *cell wall* (Fig. 2.2), which lies on the outside of the cell (plasma) membrane and consists chiefly of cellulose along with pectin or substances like lignin (wood), suberin (cork), or cutin (skin). These cell walls provide the cell and indeed the whole plant with its characteristic rigidity, and are comparable in every way (except in their chemical composition) with the formed and amorphous intercellular substances of animal tissues.

FURTHER READING

Ham A.W. (1965) *Histology* 5th ed. Chapter 10. London: Pitman Medical Publishing Co. Ltd.

Hewer E.E. (1969) *Textbook of Histology for Medical Students* 9th ed, revised by S. Bradbury. London: Heinemann.

Ramachandran G.N. (1968) *Treatise on Collagen* Vol. 2A–*The Biology of Collagen* Ed. B.S. Gould. London & New York: Academic Press.

Slavkin H.C. (1972) *The Comparative Molecular Biology of Extracellular Matrices*, New York & London: Academic Press.

4 DEVELOPMENT OF THE TISSUES–HISTOGENESIS

The development of epithelial, connective, muscle and nerve tissues is largely the story of how embryonic cells differentiate and rearrange themselves to perform special functions.

EPITHELIUM (Fig. 4.1)

This tissue covers the internal and external surfaces of the body and may be derived from ectoderm, endoderm or mesoderm. *Simple squamous epithelium*

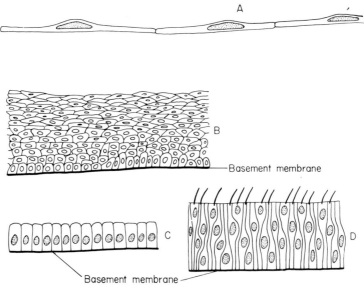

Fig. 4.1 Epithelia: (A) simple squamous; (B) stratified squamous; (C) simple columnar; (D) pseudostratified ciliated columnar.

forms a smooth delicate veneer where there is moisture, no secretion and no trauma, e.g. lining the inside of a blood vessel. The cytoplasm and organelles of the thin flattened cells are reduced to a minimum and almost the only evidence of cellular activity is provided by pinocytotic vesicles taking material across the cell in one or other direction. *Stratified squamous epithelium* protects against wear and tear. On a moist surface e.g. oesophagus, where the superficial cells of the epithelium are being continually rubbed off, they are simply replaced by proliferation of the basal cells. However, on a dry surface, e.g. skin, the cells also differentiate to produce keratin in their cytoplasm and as they reach the surface and die, this hard material which can form a remarkably thick layer not only withstands more severe trauma before it is shed but also behaves as waterproofing to prevent the loss of fluid from the underlying tissues.

When an epithelium provides a secretion or excretion or absorbs material, the cells are equipped with organelles for energy, protein synthesis and digestion or are adapted for absorption—hence the shape and size of the *columnar epithelial* cell. But the cells are easily damaged and they, too, must be replaced by mitotic division. The complete epithelial covering is also maintained with the help of modified, elaborate, desmosomes, the *junctional complexes*, where the adjacent cell walls near their exposed surfaces are partially fused (Fig. 2.1). *Cilia* are another special feature of columnar, particularly *pseudostratified columnar* epithelia, but some mystery surrounds their development.

Glands (Fig. 4.2)

When large quantities of secretion are required of an epithelium, the cells sink into the underlying tissues to form a pocket or *gland*, opening on the surface of the epithelium. A gland from which secretion escapes by a channel or duct is called *exocrine*. However, some secretions are intended for direct transfer to the bloodstream to influence distant tissues or organs; these are produced by *endocrine* glands which, developing at first like normal exocrine glands, soon lose their connections with the parent epithelium and form small groups or clusters of cells intimately associated with capillaries or sinusoids to remove the secretion. A few endocrine glands have limited storage facilities i.e. each cell cluster forms a central space or vesicle to keep secretion for emergencies or circumstances requiring increased supplies at short notice. The *basement membrane* (Fig. 4.1) which separates epithelia from the deeper tissues, consists of intercellular substances–glycoproteins and reticular fibres–products of either the epithelium or the adjacent tissues.

THE CONNECTIVE TISSUES (Fig. 4.3)

Loose connective tissue (Fig. 4.3ab)

The family of *connective tissues*–loose and dense connective tissue, cartilage and bone–are very closely related because they all have the same parent, the *primitive undifferentiated mesenchymal* cell. This small cell has numerous cytoplasmic processes extending in different directions to join up with the processes of other similar cells; the intercellular spaces in this loose mesenchyme are filled with tissue fluid. Without significant change in their appearance, these

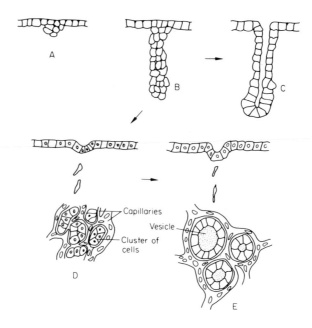

Fig. 4.2 Development of glands: (A) cells of epithelium burrowing into underlying tissue; (B) solid cord of cells proceeding to form (C), an exocrine (simple tubular) gland or (D) an endocrine gland which loses attachment to overlying epithelium and acquires network of capillaries surrounding clusters of cells; (E) development of vesicles for storing secretion of the cells (e.g. thyroid gland). After Ham, A. W. *Histology* 5th edition. Pitman Medical Publishing Co. Ltd., London.

cells become fibroblasts by synthesizing tropocollagen molecules and laying down microfibrils and then fibrils of collagen just outside their plasma membrane. A small quantity of collagen fibres closely associated with fibroblasts is characteristic of *loose connective tissue* but the same cells also produce the amorphous intercellular material, mostly hyaluronic acid, which maintains the fluid content of the tissue spaces. In short, the 'looseness' of this connective tissue depends on the paucity of collagen and the abundance of tissue fluid. The other cellular constituents of loose connective tissue are also derived from the primitive mesenchymal cells. The *macrophage*, for instance, has specialized in pinocytotic/phagocytic and lysosomic activities and *fat cells* have specialized in storing fat globules until their cytoplasm contains nothing else; adipose

Fig. 4.3 The connective tissues. (a) Loose or embryonic mesenchyme comprising primitive undifferentiated mesenchymal cells. (b) Loose connective tissue; fibroblasts surrounded by fine strands of collagen; large intercellular spaces. (c) Adipose tissue consisting chiefly of fat cells which compress the fibroblasts and collagen fibres. (d) Dense connective tissue; the fibroblasts have produced large amounts of collagen which obliterates the intercellular spaces. (e) Early stage of cartilage development (chondrogenesis); mesenchymal cells become rounded (chondroblasts) and surround themselves with collagen and amorphous intercellular substance. (f) Continued proliferation

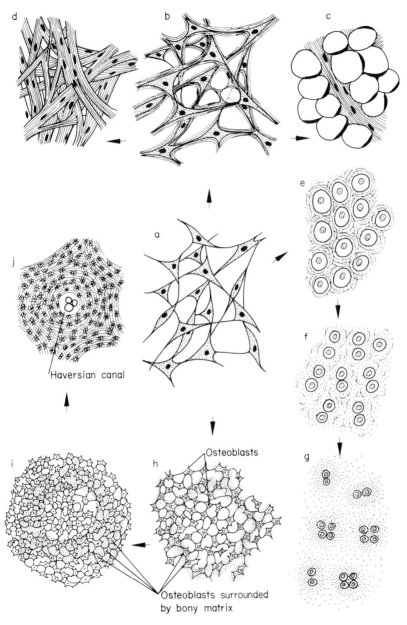

Haversian canal

Osteoblasts

Osteoblasts surrounded
by bony matrix

of chondroblasts which become separated by increasing amounts of intercellular material or matrix. (g) Fully developed cartilage. (h) Early stage of
intramembranous bone formation (osteogenesis); osteoblasts retain connections with one another and surround themselves with collagen impregnated
with calcium salts, thus forming fine strands of bone. (i) Thickening of bony
strands by involvement of increasing number of osteoblasts. (j) Adult compact bone in form of Haversian system showing concentric lamellae of bone
surrounding a Haversian canal containing blood vessels.

tissue (Fig. 4.3c) is no more than loose connective tissue with an exceptionally large number of fat cells in it.

Dense connective tissue (Fig. 4.3d)

This tissue appears when pressure or tension is applied to connective tissue and where support or resistance is required; the fibroblasts simply produce more and more collagen until most of the intercellular spaces are filled with large, irregularly arranged collagen bundles. The amorphous intercellular substances behave like an embedding jelly-like material for the collagen and the concentration of sulphated mucopolysaccharide (e.g. chondroitin sulphate) is increased to provide a firmer consistence. Orientation of the collagen bundles depends on the forces playing on the tissue; in skin, where no consistent directional tension is exerted, the collagen forms an irregular meshwork but in tendon, where tension is always along one axis, the collagen is arranged in regular and parallel bundles.

Cartilage (Fig. 4.3e–g)

Developmentally this can be regarded as merely a further modification of the elements found in loose and dense connective tissue; as soon as the mesenchymal cells begin to produce collagen and chondroitin sulphate they withdraw their cytoplasmic processes and become rounded. Thus each cell surrounds itself with matrix, i.e. closely packed bundles of collagen embedded in chondroitin sulphate. All intercellular spaces disappear and the mass becomes rubbery hard and homogeneous or *hyaline* in appearance. Further growth of the cartilaginous mass is always necessary; thus, early in development, each cartilage-forming cell (the chondroblast) continues to undergo mitosis and each daughter cell forces itself apart from its neighbour by surrounding itself with matrix. This *interstitial growth* continues for a limited period only and evidence of it remains in the form of cell nests—groups of two, three or four cartilage cells separated by slender septa of matrix—all that can be produced in the face of increasing pressure. But cartilage must continue to grow and it does so by *appositional* (or *perichondrial*) *growth*; perichondrium, the sheath of tissue surrounding a mass of cartilage consists of cells graded in shape and relationships from the spindle-shaped fibroblasts at the periphery to the rounded chondroblasts embedded in the cartilage. During appositional growth there is proliferation of the fibroblasts in the perichondrium followed by their transformation into chondroblasts and a progressive deposition of cartilage until the mass reaches its full size. Cartilage acquires its blood supply by engulfing adjacent blood vessels as it grows; and since the blood vessels are not always in close apposition to all its cells, nutrition must be maintained by diffusion through the matrix. In some parts of the body, e.g. external ear, larynx, the chondroblasts synthesize elastic fibres which permeate the matrix and provide a degree of elastic recoil (elastic cartilage). In other regions, e.g. intervertebral disc, where tension as well as pressure is exerted on the tissue, the matrix contains large quantities of dense connective tissue rather than consisting entirely of hyaline cartilage (fibrocartilage).

Bone (Figs. 4.3h–j, 4.4)

Bone was designed and instituted during evolution when a stronger and more rigid form of support than cartilage became necessary and no doubt, at some

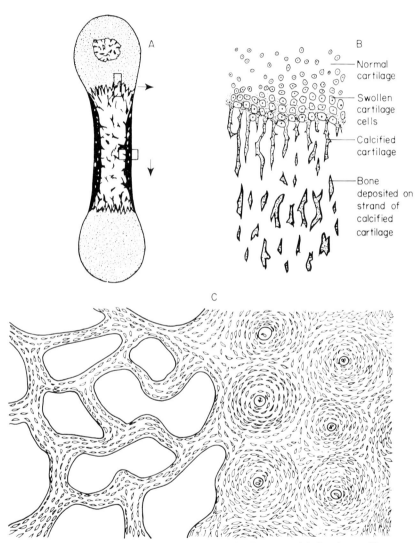

Fig. 4.4 Endochondral ossification. (A) Cartilaginous model showing ossification of shaft and a secondary ossification centre at upper end. (B) Section through the area of the transition from cartilage to bone. (C) (Left) Cancellous or spongy bone comprising strands of bone and large spaces containing blood vessels and bone marrow; (Right) Compact bone formed by continued deposition of bone reducing the vascular spaces to Haversian canals.

stage, attempts were made to obtain this kind of support by simply calcifying the cartilage already present. Evidence of these abortive measures and the reason for their failure can be seen during *endochondral ossification.* Many bones, e.g. those of the limbs, are first represented by small cartilaginous models which continue to increase in size by interstitial growth. The first sign of bone occurs in mid shaft—at the *primary centre of ossification* (Fig. 4.4A). The chondrocytes there become swollen and bloated, compressing the intervening matrix and depositing calcium salts (*calcification*) in it. This process spreads in both directions along the shaft, involving more and more chondrocytes. Eventually these cells find themselves sequestered in a honeycomb of calcified cartilage, but unfortunately, because diffusion is not possible through calcified cartilage, they are then cut off from their source of nutrition. These chondrocytes therefore die, leaving calcified strands which are also in danger of disintegrating after the death of the cells. Meanwhile, on the surface of the shaft, the cells of the vascular perichondrium deposit a sleeve of periosteal bone, i.e. they become *osteoblasts,* which not only synthesize and envelop themselves in collagen but also, like the bloated cartilage cells in the centre of the shaft, deposit insoluble calcium salts on these fibres. Unlike chondroblasts, however, osteoblasts retain their proto-plasmic processes and therefore their links with other osteoblasts and, through these, with the uncommitted mesenchymal cells and blood vessels outside the bone-forming area (Fig. 4.3h—i). But blood vessels also become incorporated in this periosteal sleeve of bone and, through the minute canaliculi containing the protoplasmic processes, the osteoblasts, even when deeply buried in a mass of bone, can obtain nourishment by relay from these vessels. From these same blood vessels in this periosteal bone, tiny capillaries, accompanied by undifferen-tiated mesenchymal cells now invade the shaft and enter the spaces originally occupied by chondroblasts between the calcified strands. The invading cells become osteoblasts which proceed to form layers of bone on the surfaces of the calcified cartilage. The result is an irregular meshwork of bony strands, i.e. *cancellous bone,* occupying the whole width of the shaft. But this is a temporary measure. At the periphery, many of the strands will become thicker and form the cortex (see below) but, centrally, they are absorbed and removed by the action of *osteoclasts* (bone-destroying cells)—also differentiated from the mesenchymal cells; the medullary cavity which is left becomes occupied by vascular tissue, the precursor of bone-marrow.

Cancellous or spongy bone consists of more spaces than bony strands but the cortex consists of *compact bone* (Fig. 4.4C); this is obtained by continuing cancellous bone development, i.e. deposition of bone goes on, layer upon layer, on the bony strands forming the walls of the spaces, until the spaces themselves become reduced to very narrow channels containing only the blood vessels. This arrangement of concentric layers of bone surrounding a narrow vascular channel (Fig. 4.3j, 4.4C), is a *Haversian system* and compact bone consists entirely of these systems. The blood vessels of the Haversian canal, can sustain all the osteocytes within its own system through the meshwork of lacunae and canaliculi permeating the bone.

While ossification is spreading from the primary centre towards the ends of the shaft, the cartilaginous model continues to grow in length and in nearly every long bone, a *secondary centre* of ossification appears in the cartilaginous knob at each end (Fig. 4.4). As this new centre increases in size, a layer of cartilage, the *epiphyseal plate*, between the shaft (diaphysis) and the end (epiphysis) of the bone, finds itself in the uncomfortable situation of being 'invaded' by an ossification process on each side. By means of vigorous interstitial growth it continues to produce cartilage for the advancing endochondral processes of ossification on either side. Gradually, ossification overtakes cartilaginous growth and when the two processes of ossification meet, bone growth ceases. It is worth mentioning here that the process whereby chondrocytes and osteoblasts can cause the deposition of insoluble calcium salts is still not fully understood. Perhaps it is enough to suggest that these cells can synthesize alkaline phosphatase which releases phosphate ions from phosphorylated sugars in the surrounding tissue fluids: the calcium/phosphate balance in the vicinity is thereby disturbed and, because of the increased concentration of phosphate ions, complex insoluble calcium phosphate salts are precipitated in the surrounding matrix.

Bones such as the mandible and maxilla which develop by *intra-membranous ossification* (Fig. 4.3) have a much less complicated life history. Some of the stellate mesenchymal cells, which retain their protoplasmic processes and hence their links with one another, surround themselves with fine sheaths of bone i.e. collagen fibres impregnated with calcium salts. The intervening spaces or channels contain blood vessels and uncommitted mesenchymal cells, the whole picture closely resembling that of the early stages of spongy bone development by the endochondral process. With thickening of some bony strands by layers of bone and osteoblasts, and absorption of others by osteoclasts, the tissue acquires the characteristics of spongy bone; however, progressive deposition of bone at the expense of the vascular spaces gives *compact bone* typified, as in endochondral ossification, by Haversian systems with their concentric bony lamellae and central vascular canals.

Haemopoietic tissues

The argument for including *blood* and the *haemopoietic tissues* with the connective tissues is not being defended here but it is true that their development is more easily understood when described along with the connective tissues. Morphologically, blood has the same features i.e. cells and intercellular substance, the latter normally in a fluid phase, the plasma; under some circumstances, however, owing to changes in its chemistry and physical properties, this intercellular fluid clots or coagulates becoming a very firm material in which the cells are trapped. The development of the haemopoietic tissues is yet another example of the potentiality of the primitive mesenchymal cell.

In the wide spaces between the strands of early spongy bone, some of the primitive mesenchymal cells are diverted (as osteoblasts) towards the strengthening of the bony framework but others become reticulo-endothelial cells.

These large oval cells have two functions; they may form fine reticular fibres
to which they cling and thus produce a delicate meshwork to support the blood-
forming tissue (Fig. 4.5); but many of the cells arrange themselves to form an

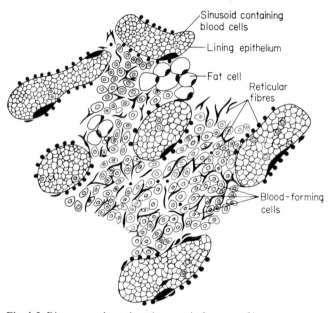

Fig. 4.5 Diagram to show the microscopic features of bone marrow.

incomplete lining for the wide sinusoids of the bone marrow leaving many
spaces for blood cells to pass freely through the walls. Outside the sinusoids,
in the meshes of the reticular framework, masses of cells differentiate into the
various types of blood cells. They can all trace their ancestry back to the
primitive mesenchymal cells but the first step along each line appears,
morphologically, to be the same cell, the *myeloblast* or *haemocytoblast* (Fig. 4.6).
This large round cell differentiates or matures smoothly through numerous
identifiable stages to become *red* or *white blood cells*; Mitoses are frequent,
with some daughter cells always remaining as a reservoir for further cell
divisions, while the rest continue to the mature forms.

During development of the *red blood cell* (Fig. 4.6) each stage sees a gradual
reduction in the size of the cell; the nucleus also shrinks until it becomes a dense
compact mass of chromatin; the cytoplasm assumes a darker shade of blue
because of the increased number of ribosomes. As the haemoglobin is synthesized
and accumulates, the cell stains with blue *and* red dyes (the *polychromatic
erythroblast*) until, with the complete takeover of the cytoplasm by haemoglobin,
the basophilic material disappears (*normoblast*). Meanwhile, the nucleus, its
task complete, is discarded by the cell (*erythrocyte*) which becomes an inert
membranous sac containing only haemoglobin. Even in birds and amphibia,
where the nucleus is retained in the erythrocyte, its small densely-stained
appearance indicates that it is no longer functional.

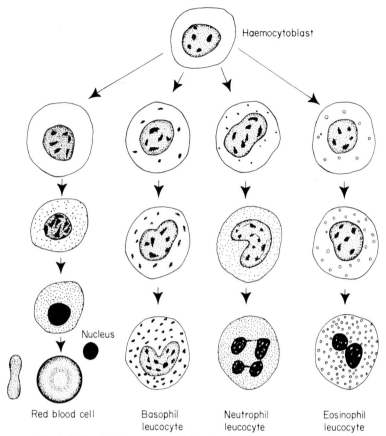

Haemocytoblast

Nucleus

Red blood cell

Basophil
leucocyte

Neutrophil
leucocyte

Eosinophil
leucocyte

Fig. 4.6 The development of red blood cells and granular leucocytes.

For the *white blood cells* (Fig. 4.6) there is the same smooth progression from the myeloblast along one of three different lines to give a *neutrophil, eosinophil* or *basophil leucocyte*. The *promyelocyte*, a slightly smaller cell than the myeloblast contains a few nondescript granules in its cytoplasm and, not until the *myelocyte* stage, are these granules sufficiently prominent or numerous to indicate by their staining reaction whether the line is neutrophil, eosinophil or basophil; the nucleus, so far, has only a slight indentation. In the *leucocyte*, granules reach the adult concentration but, as it matures further, first in the marrow and then in the blood stream, there is both a gradual diminution in cell size and a progressive lobation of the nucleus, more marked in the case of the neutrophil, which may have a five-lobed nucleus, and least marked in the eosinophil which seldom has more than two lobes. The precise nature of the granules characterizing the leucocytes is still in some doubt but there is now good evidence that they are, in fact, lysosomes, a view which is consistent with the recognized function of these cells, viz. the ingestion and destruction of bacteria. In the *megakaryocyte*, a very large cell anchored to the reticular framework of the bone marrow, the nucleus is a mass of large round lobes partially overlapping one another; the cytoplasm, after the development of a system of intracytoplasmic membranes, breaks up into fragments which escape into the blood as *platelets*. No one has been able to demonstrate clearly how *monocytes* arise. In the early fetus, blood cells are also produced in the liver and spleen but at birth these organs seldom show signs of haemopoiesis. Later, most of the myeloid tissue in the shaft of a bone ceases to produce blood cells and becomes dormant, turning a yellow colour because the marrow is populated with fat cells. The marrow at the ends of a bone continue throughout life to produce blood cells. If and when necessary, e.g. after severe and repeated blood loss, the yellow marrow may change to red marrow—evidence that stem cells, hidden among the fat cells, are still present and capable of differentiation and maturation into blood cells.

Like myeloid tissue, *lymphoid* or lymphatic tissue develops from mesenchyme by differentiation of the primitive mesenchymal cells (a) into reticulo-endothelial cells, which form and adhere to a reticular framework and (b) into lymphoblasts, precursors of the lymphocytes. Since lymphocytes are prominent in the defence of the body against invasion and infection by micro-organisms, the lymphoid tissue develops where there is a risk of such invasion and is strategically placed to intercept it. Thus, under the moist epithelium of the alimentary, urinary and upper respiratory tracts, white *primary nodules* (Fig. 4.7) actively producing or ready to produce lymphocytes are scattered throughout the loose connective tissue. In the reticular meshwork of these nodules lie the *lymphoblasts*, slightly smaller than the parent mesemchymal cell and containing an appreciable content of pale blue cytoplasm. Further maturation produces the *prolympho-cytes* with small dense nuclei and enough cytoplasm to be obvious. This cell, or one very like it, is the *large lymphocyte* of the blood stream. The typical *small lymphocyte* of the blood is also present in the nodules. In a stained section the nodule appears under low power as a dark blue almost black mass, because there are so many closely packed lymphocytes each with its small dense nucleus and only a minimum of cytoplasm. With increased lymphocyte production however, the centre of the nodule becomes much paler owing to the intense proliferation there of the precursor and early lymphoid cells (lymphoblasts) which have large 'open' nuclei and a considerable volume of pale blue cytoplasm; this is the *germinal centre* (Fig. 4.7). The end products i.e. small lymphocytes, are pushed towards the periphery to give the nodule a densely stained rim. The more organized lymphoid masses, e.g. *tonsils, adenoids, lymph nodes, white pulp of spleen*, develop in the same way. The development of the *thymus gland*, also an important lymphoid organ, is described on page 79.

MUSCLE

Although three distinct and different types of *muscle* are recognized—*smooth*, *skeletal* (*striated*) and *cardiac*—all are developed, along with the connective

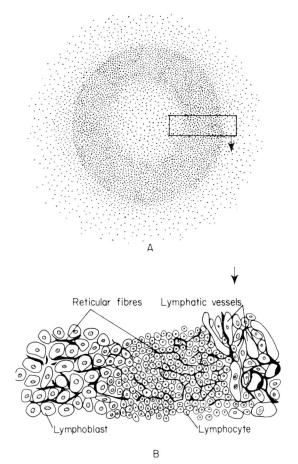

Reticular fibres   Lymphatic vessels

Lymphoblast   Lymphocyte

B

Fig. 4.7 (A) Nodule of lymphoid tissue showing pale germinal centre. (B) Higher magnification of area outlined in (A) showing reticular fibres supporting early lymphoid cells, developing lymphocytes and the mature lymphocytes ready to enter the lymphatic vessels.

tissues, from the early mesenchymal cells of the embryo. Connective tissue and muscle are probably derived from the same parent cells but it is extremely difficult except with special techniques to tell when a primitive mesenchymal cell takes its first step towards becoming a myoblast, i.e. the synthesis of contractile proteins in the form of myofibrils in its cytoplasm.

Smooth muscle (Fig. 4.8B)

The individual spindle-shaped *smooth muscle cells* or *fibres* have their myofibrils arranged along the long axis of the cell with the nucleus in the centre. The cells are separated from, as well as anchored to, one another by fine collagenous or reticular fibres but the source of this extracellular material

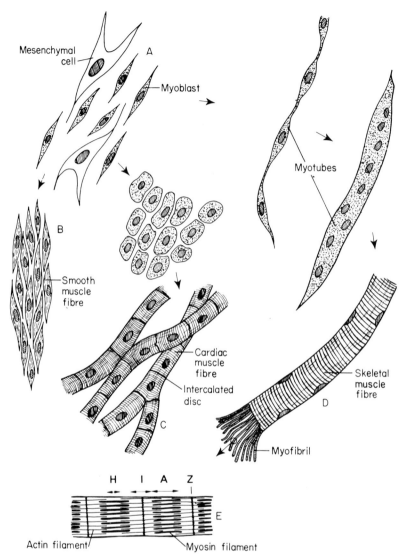

Fig. 4.8 The development of smooth muscle (B), cardiac muscle fibres (C) and skeletal muscle fibres (D) from mesenchyme (A). (E) The fine structure of the myofibril as seen with the electron microscope.

is uncertain—either from the muscle cell itself or (more likely) from concomitant fibroblasts.

Skeletal muscle (Fig. 4.8d,e)

The histogenesis of skeletal muscle is more intriguing; the adult fibre, a long multinucleate syncytial structure is produced by end to end fusion of many separate myoblasts until a *myotube* is formed and further growth in length occurs by the addition of fresh myoblasts to the surface and to either end. The enormous increase in the number of actin and myosin filaments leads to the displacement of the nuclei to the periphery of the fibre but the organisation of these filaments into typical myofibrils with their characteristic sarcomeres and cross-striations is not yet fully understood. Mitochondria accumulate in large numbers around the myofibrils while fine cytoplasmic tubules, responsible for the conduction of the nerve impulse, grow inwards from the sarcolemma to surround the myofibrils. How the myoblasts are first stimulated to differentiate and to form myotubes is also unknown, but these developments can take place in the absence of nerve fibres; however fibroblasts, or at least their product, collagen, seem to play an essential role in the differentiation of myoblasts and myotubes.

Cardiac muscle (Fig. 4.8c)

Like skeletal muscle, development of the *cardiac muscle fibre* occurs as the result of end-to-end apposition of myoblasts to one another but, unlike skeletal muscle, the opposing cell membranes do not disappear; indeed they thicken to form desmosome-like adhesions with filamentous thickenings radiating into the cytoplasm. The irregular interdigitations of these cell junctions crossing the long axis of the fibre are the intercalated discs characteristic of cardiac muscle.

NERVE TISSUE (Fig. 4.9)

Nerve tissue consists of nerve cells or neurones and a 'connective', *supporting* or *sustentacular* tissue component which varies according to the site of the neurone and with the part of the neurone involved.

All that concerns us here is the development of the neurone and its supporting tissue, i.e. the histogenesis of the nervous tissue, leaving its morphogenesis until the other systems are dealt with (chapter 7). The neural tube and neural crest (p.31) are the sites of development of nerve cells and their supporting tissue. In the wall of the neural tube (Fig. 4.10A–C) the cells initially form a simple columnar epithelium; later they elongate to form a pseudo-stratified epithelium. Some of these cells lose their attachments to the outer and inner surfaces of the tube, become free in the mid-zone (mantle layer) of the wall and develop into nerve cells or neuroblasts. At first these cells have only two protoplasmic processes; one process continues to increase in length to become the axon while the other develops into a number of branching dendrites projecting from a wide area of the cell surface. The former, the axon, may seek out and gain attachment to another cell body or to the end of a dendrite from another cell, forming a synapse (see below) or it may grow out of the neural tube to reach a developing muscle or gland; in either case, it carries an impulse *from* the nerve cell whereas the function of the dendrites is to collect and carry impulses *to* the nerve cell.

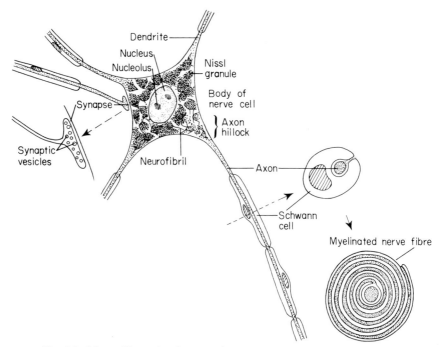

Fig. 4.9 Scheme illustrating features of a neurone, a synapse and myelination.

The cytological details of development and differentiation in the neurone are merely elaborations of the features in a basic cell. The cytoplasm accumulates enormous quantities of thin collapsed rough-surfaced vesicles of the endoplasmic reticulum; the localized concentrations of these, the *Nissl substance* or *granules* (Fig. 4.9), were identified many years ago. Although rough-surfaced vesicles of this nature are an indication of protein synthesis in a cell, the nature of that material in the nerve cell has not been identified. Fine neurofilaments appear, permeating the whole cytoplasm and extending the length of each process: these filaments can only be distinguished individually by the EM but in ordinary histological preparations, they become aggregated and identifiable as the *neurofibrils*. A *synapse* (Fig. 4.9) obviously results from apposition of the processes of two different cells or of an axon to the body of another cell but there is no certainty about how these areas of contact are chosen or persist. Somewhat similar in structure as well as in function are the *neuromuscular junctions* where an axonal ending is in apposition to the sarcolemma of a muscle cell. In both cases, once they have been established, the contacts are maintained, throughout growth and development, by elongation of the nerve cell processes, in spite of the increasing distance between the two nerve cells or between the body of the nerve cell and the muscle fibre.

Protection and perhaps even the insulation of the nerve cell and its processes by the supporting tissue are essential—the simplest example being the *neurolemmal (Schwann cell) sheath* (Fig. 4.9) around the peripheral nerve fibre. After

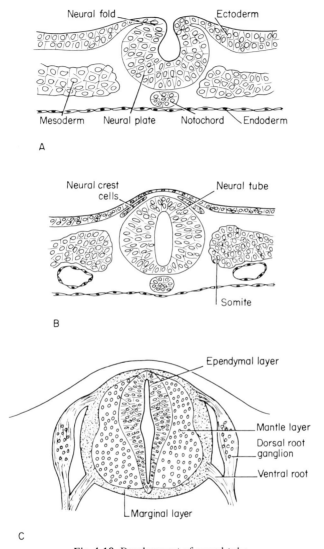

Fig. 4.10 Development of neural tube.

the nerve fibre has emerged from the spinal cord or ganglion it is joined by
Schwann cells, which probably originate from the neural crest. Spaced along the
fibre or fibres, these cells elongate and allow the fibres to sink into their
cytoplasm still surrounded by a covering of cell membrane and connected to the
surface of the cell by the mesaxon, i.e. the two layers of the cell membrane
reaching back from the nerve fibre to the cell surface. In this way the whole
length of a nerve fibre is embedded, segment by segment, in the cytoplasm of
Schwann cells. These fibres are non-myelinated. This is only a developmental
stage for many fibres as they will later become myelinated, i.e. covered by

myelin. Myelin itself consists of innumerable layers of cell membrane wrapped closely round the fibre and it is generally believed that this arrangement of membranes is obtained by the mesaxon wrapping itself repeatedly round the fibre or by the rotation of the fibre dragging the mesaxon around it like a roller blind. Instead of this almost mechanical concept, it has been suggested that the layers are derived from other less well defined cytoplasmic membranes which are somehow organized and intercalated into the regular pattern of the typical myelin sheath.

Within the central nervous system, the Schwann cells are replaced by the oligodendrocytes. The development of myelin within these cells is believed to occur by the last method described above. The other supporting cells in the brain and spinal cord, i.e. astrocytes and ependymal cells, arise along with the oligodendrocytes from the epithelial-like cells of the neural tube and are known as the *neuroglia*. The *microglial* cells which appear later amongst the nerve cells are thought to be derived from mesenchyme. In peripheral ganglia, the *capsule* or *satellite cells* surrounding the nerve cells are derived from the neural crest.

FURTHER READING

Banker B.Q. Przybylski R.J. Van Der Meulen J.P. & Victor M. (1972) *Research in Muscle Development and the Muscle Spindle.* Amsterdam: Excerpta Medica.

Bourne G.H. (1971) *The Biochemistry and Physiology of Bone* 2nd ed.Vol. III. New York & London: Academic Press.

Bourne G.H. (1973) *The Structure and Function of Muscle* 2nd ed.Vol. I. New York & London: Academic Press.

Freeman W.H. & Bracegirdle B. (1967) *An Atlas of Histology* 2nd ed. London: Heinemann Educational Books Ltd.

Ham A.W. (1965) *Histology* 5th ed Chapters 8–21. London: Pitman Publishing Co. Ltd.

Mauro A. Shafiq S.A. & Milhorat A.T. (1970) *Regeneration of Striated Muscle and Myogenesis.* Amsterdam: Excerpta Medica.

Potter K.R. & Bonneville M.A. (1968) *Fine Structure of Cells and Tissues* 3rd ed. Philadelphia: Lea & Febiger.

The Open University (1974) *Physiology of Cells and Organisms* Units 5 & 6. Milton Keynes: The Open University Press.

Windle W.F. (1960) *Textbook of Histology* 3rd ed. Chapters 1–8, 11 & 12. New York & London: McGraw-Hill Book Company, Inc.

5 THE EARLY STAGES OF DEVELOPMENT

If the development of every creature from fertilization through to its adult form were to be described, the result would be an enormous treatise. We must treat comparative development in a more rational way by finding and defining a *basic* pattern; it should then be possible to see the modification(s) of this pattern in the development of each animal, thus demonstrating the evolution of development as well as the factors involved in evolution. Instead of beginning with the earliest invertebrates, working through them to the vertebrates and dealing exhaustively with these as far as the human, we shall choose only enough examples to illustrate the concept and the factors modifying the pattern. The amphioxus, a primitive chordate standing close to the transition from invertebrates to vertebrates, may seem too far advanced to illustrate the variations in development among the invertebrates but it does form the most convenient starting point; it is easy, for instance, to see in it the basic plan which is modified in the vertebrates; and these modifications are best illustrated by choosing an amphibian, then a bird or reptile and finally a mammal.

THE PRIMITIVE BODY FORM (Fig. 5.1)

But the aim or end-point of every developmental process, i.e. the adult form, is also important. Unfortunately adult animals vary so greatly in their final form that some sort of basic plan is necessary here as well in order to make comparisons. Thus we have to define those features which are common to every animal from the primitive chordate through to the human and then assemble them as a basic or *primitive body form* which can be visualized as follows: a sausage-shaped entity with an outer tube or envelope as the skin; inside is a second tube representing the alimentary canal; the two are continuous with one another at either end, forming the *mouth* in front and the *anus* behind. The space between the tubes contains the other features: the *neural tube*, or rudimentary nervous system lies below the dorsal surface and extends from one end to the other. Ventral to the neural tube is a firm strand of tissue, the *notochord*, serving as a stiffening rod for the whole body. *Mesodermal tissue*, like a packing material round the neural tube and notochord, fills the rest of the

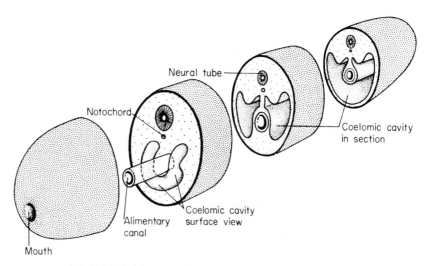

Fig. 5.1 Primitive body form sectioned to show main features.

space but has a fluid-filled cavity, the *coelom*, almost encircling the alimentary canal. Mesoderm differentiates into muscle, cartilage and bone as well as the blood, blood vessels and heart. Projections from the surface of the adult body, e.g. limbs, ears and nose, are merely localized proliferations and specializations of mesoderm. The primitive body form is important because every embryo passes through this stage; it may not be—and rarely is—arranged exactly as shown in Fig. 5.1, nor do all the elements of the primitive body form develop in the same sequence in all animals.

Eggs or ova are of many different kinds and the next exercise in tracing the development of any creature is to discover how an egg reaches the primitive body form; fortunately the amphioxus, and amphibian, a bird and a mammal also provide enough variations in size of ovum and other characteristics to establish a pattern of development which can be extrapolated to form a plan for the invertebrates and also be adapted for all the intermediate evolutionary steps among the vertebrates.

Development beyond the primitive body form is quite different. With all the basic features present by this stage, it is more logical to describe the general plan of development for each organ and system and, while doing so, to refer to the variations occurring in different species. The scheme is shown below.

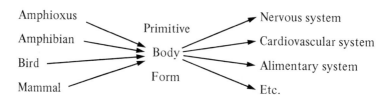

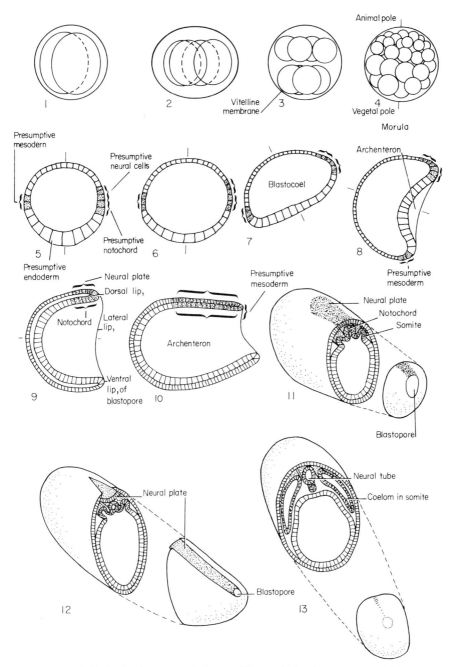

Fig. 5.2 Early development of the amphioxus. 1–4, cleavage to morula; 5–10, blastula through gastrulation (in section); 11–13, tubulation.

AMPHIOXUS (Fig. 5.2)

The phenomena which follow fertilization (rather than fertilization itself) offer a convenient if not a more logical starting point for describing the development of an animal and, as we indicated earlier, the best organism to start with is the amphioxus; its ovum has a diameter of only 0.1mm. Within the *zygote* or fertilized ovum, its own nucleus (the *female pronucleus*) and the nucleus of the spermatozoon (*male pronucleus*) fuse by a process similar to mitosis; thus, from each pronucleus the chromosomes, tightly coiled and easily stained, are released into the cytoplasm, only to assemble in one mitotic spindle; thereafter it is mitotic division with the formation of two nuclei and two new cells; This is the first of a series of cell divisions known collectively as *cleavage*. The first cell division takes place along a *meridional plane*, i.e. passing through both the 'north' and 'south' poles, and the next, which involves the two new cells, is also meridional but at right angles to the first; the result is four cells. Each of these now divides but their plane of cleavage is equatorial or nearly so; in fact, it is usually slightly above the equator because of uneven distribution of the *yolk* or food in the cytoplasm; this inert material usually gravitates towards the lower, *vegetal*, pole displacing many of the cytoplasmic organelles from that region; conversely, the opposite, *animal*, pole, has less yolk and more organelles, becoming the more active part of the ovum. Consequently the nucleus lies nearer the animal pole and the plane of the equatorial cleavage is above the true equator. Therefore, at the end of the third cleavage, four small cells, the *micromeres* sit on top of four larger cells, the *macromeres*. From now onwards the rate of cleavage in the more metabolically active micromeres is greater than in the macromeres and, after several cleavages, there are many more micromeres than macromeres; together they form a spherical mass of cells called the *morula*. Next, a fluid-filled cavity, the blastocoel, appears in the middle of the mass which, until now, was comfortably enclosed in an envelope called the *vitelline membrane* surrounding the ovum from the beginning. With increasing volume of the mass, the vitelline membrane disintegrates, and as the fluid in the blastocoel accumulates, the cells, micromeres and macromeres, form a single layer around the cavity. This is the *blastula* stage.

Perhaps as a result of the slower cleavage rate among the macromeres, a flattening of the vegetal pole occurs until the blastula resembles a hemisphere rather than a sphere. But the macromeres do not remain as a flat plate; they become drawn into the hollow dome of micromeres, reducing the size of the blastocoel and forming a kind of double-walled cup. The process looks exactly like the collapse of a punctured rubber ball and the shallow depression thus formed becomes the *archenteron*, opening widely to the outside. The opening is the *blastopore* and its edge, where the inner and outer layers meet, forms the *lips* of the blastopore. The blastocoel, meanwhile, disappears. At this *early gastrula* stage, the orientation of the whole mass changes; it gradually tilts until the original animal pole becomes the front end of the gastrula, and the blastopore looks backwards. The most important phases of *gastrulation* now follow and the cells at the lips of the blastopore perform the most active role. That part of the blastoporal lip lying uppermost in the new position of the gastrula is the *dorsal lip*; opposite is the *ventral lip* and on each side a *lateral lip*. In essence, the cells around the blastopore proliferate, and add to the walls of the cup (*tubulation*) until the archenteron is a tube-like cavity ending blindly at the head end but opening backwards through the blastopore which becomes progressively narrower. Generally, the cells of the outer layer of the gastrula form the ectoderm or outer covering of the embryo, and the cells of the inner layer provide the mesoderm and endoderm. However, at the ventral lip and adjoining parts of the lateral lips, the cells simply increase the length of the archenteron by adding to the inner and outer layers. In doing so, they outstrip the other cells around the blastopore and the result is a tilting of the blastopore so that it faces dorsally as well as caudally and, in fact, it eventually opens on the dorsum of the gastrula. The activity of the cells around the rest of the blastoporal lip is more complicated. By using special stains which do not harm the cells, we learn that there is a migration of some cells from the ventral lip around each side of the blastopore to a position high up on the lateral lip; by the same technique, two groups of cells can be identified in the dorsal lip. These four groups of cells all proliferate and add to the length of the gastrula in their own way. Thus, the outer cells in the dorsal lip (the presumptive neural cells) lay down in the ecto-derm an elongating band of cells (the neural plate) along the midline of the dorsum; to begin with, this plate is indistinguishable morphologically from the adjacent ectoderm. At the same time and in the same fashion, the inner group of cells proliferate and leave a narrow band the *presumptive notochord* in the midline of the roof of the archenteron. On each side of this notochordal band, the cells which have migrated from the ventral to the lateral lip, the *presumptive mesoderm*, likewise contribute a band of cells to the roof of the archenteron. These presumptive tissues have still to reach their definitive positions; the term 'presumptive' is given to these groups of cells at these early stages because we can presume that under normal conditions of migration, development and differentiation, they will later form those particular adult tissues. Starting at the cephalic end and progressing caudally, the two edges of the notochordal band begin to curl up forming a groove opening towards the archenteron. As the edges come together and fuse, a hollow tube of notochordal cells sinks into the space between ectoderm and endoderm, and the roof of the archenteron heals up, The tube soon loses its lumen and becomes a solid notochord.

Along the edges of the neural plate, the adjacent ectoderm grows over the plate as a hood or cover, separated from it by a distinct space. Later, the edges of the neural plate curl up in the opposite direction to that of the notochordal plate i.e. towards the dorsum of the embryo, forming the definitive *neural tube* but, unlike the notochord, retaining its lumen.

The band of mesoderm on each side of the presumptive notochord behaves like the presumptive notochord but, instead of its edges rising up to form a tube, it behaves as if it were a series of segments, each segment, separately, invaginating the space between ectoderm and endoderm to form a vesicle which finally separates from the endoderm. With the departure of presumptive notochordal and mesodermal tissue from its roof, the archenteron is entirely walled by endoderm. There now lies on either side of the notochord a series of mesodermal vesicles called *somites* which increase in size and grow round the body in the space between ectoderm and endoderm. The cavities of the somites are retained and although they remain separate for some time, they finally join up to provide a continuous internal cavity lined with mesoderm called the *coelom*. The primitive mesenchymal tissue, derived from the walls of the somites, differentiates into muscle, skeleton, connective tissue, blood and blood vessels.

The development of a mouth obviously requires the breakdown of ectoderm and underlying endoderm in an area near the head end of the embryo. The first sign of this (and indeed an essential for the breakdown of any epithelial layers) is the absence of mesoderm between ectoderm and endoderm. In some of the invertebrates, the blastopore is a ready-made anus and fulfils that function through to adult life but in the amphioxus and all vertebrates, the anus is formed in a fashion similar to that of the mouth in an area just ventral to the blastopore which eventually closes and disappears without trace.

Although features like presumptive notochord, presumptive mesoderm and so on were noted in the gastrula as groups of cells which gave rise later to definitive structures and tissues, they can be distinguished by vital staining at even earlier stages e.g. in the morula and, surprisingly, in the ovum as areas of cytoplasm. The delineations and representation of developing and differentiating tissues on the surface of zygote and embryo through all its different stages are known as (*presumptive*) *fate maps*.

AMPHIBIAN—THE FROG (Fig. 5.3)

The amphibian has a much larger ovum than amphioxus—2mm in diameter—because the cytoplasm contains more yolk. The increased content of yolk allows for a longer period of development and therefore a more advanced and complex organism, better equipped to fend for itself in finding its own nourishment from the environment; and the presence of so much yolk, is largely, if not entirely, responsible for the modifications in the pattern of gastrulation in the frog.

The pattern which the amphibia inherited was a fundamental or basic feature and could not be discarded; they could only adapt or modify it to suit any newly acquired characteristics such as the great increase in food stores in the cytoplasm; and these stores, like those of the amphioxus, are unevenly distributed, concentrated in the vegetal pole and creating an even greater differential in the distribution of active cytoplasmic organelles between animal and vegetal poles.

After fertilization, when the presence of the first two diploid nuclei initiate

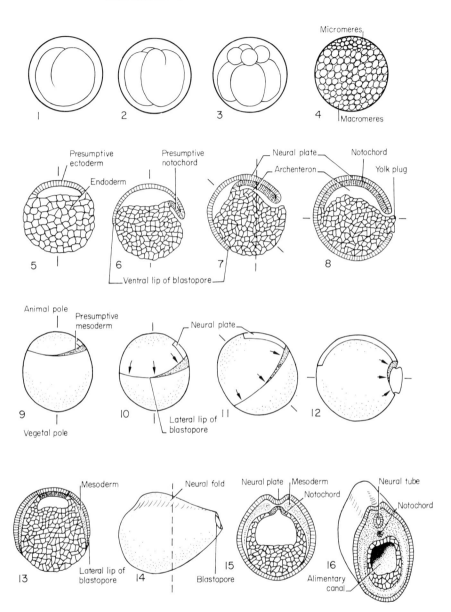

Fig. 5.3 Early development of an amphibian. 1–4, cleavage to morula stages; 5–8, sections of, and 9–12, surface views of morula and gastrulation stages; 13, transverse section as indicated in 7; 14, neurula, side view; 15, section as indicated in 14. 16, neurulation completed.

the first cleavage, the furrow between the two daughter cells appears over the animal pole but its progress towards the vegetal pole is impeded by the heavy concentration of yolk. It gets there in the end and two distinct cells are formed but not before the next cleavage has begun at the animal pole and produced a second furrow at right angles to the first. This, too, makes its way slowly towards the opposite pole. And, because of the unequal distribution of yolk, the position of the third or 'equatorial' cleavage is even further removed from the equator than in the amphioxus, giving a greater discrepancy between the resultant micromeres and macromeres. As cleavage progresses through the morula stage, this becomes even more obvious, and the number of micromeres soon outstrips that of macromeres. In the blastula the fluid-filled blastocoel is small and confined to the region of the animal pole.

Thus far, the yolk has not proved an insuperable difficulty to the embryo in following the inherited pattern of development but the plan now requires the presumptive endoderm i.e. the macromeres, to be invaginated within the dome of micromeres—virtually impossible because of the enormous size of the macromere mass and the relatively small micromere dome. The simple solution is for the micromeres i.e. the presumptive ectoderm to proliferate and grow down over the surface of the macromeres enclosing them in a coat of ectoderm, a process known as *epiboly*. Although that is essentially what happens, the process is not as simple as it sounds. Presumptive mesoderm, presumptive neural tissue and particularly the presumptive notochord (the important organizing area) all play a part. The fate map of the frog blastula (Fig. 5.3) shows that the positions of all the presumptive tissues are like those of the amphioxus, except that the presumptive mesoderm already lies close to the presumptive notochord. The importance of invagination during gastrulation is not 'forgotten' in spite of the overwhelming odds against it imposed by the additional yolk; and furthermore an archenteron or gut is still essential. Therefore, whilst proliferation of the presumptive ectodermal cells extends their territory and thereby covers the endoderm, the presumptive notochordal cells situated on one side at the edge of the ectodermal dome turn inwards or invaginate into the mass of the blastula. In doing so, they and the presumptive neural cells occupy the same positions as they do in the dorsal lip of the blastopore in the amphioxus. This *is* the dorsal lip of the amphibian blastopore and it develops as in the amphioxus, by proliferation of its cells, growing down over the rest of the gastrula towards the vegetal pole and leaving an elongated plaque of neural cells in the ectodermal layer and a similar plaque of notochordal cells immediately underneath and closely applied to the neural plate. This notochordal plate is separated from the endodermal (yolk-laden) cells by a new space—the archenteron. Meantime the rest of the sheet of presumptive ectodermal cells, spreading over the endodermal cells, has an edge which represents the two lateral lips and ventral lip of the blastopore. All four lips are therefore proceeding towards the vegetal pole encircling a blastopore plugged by endodermal cells. Yet, examination of the blastopore throughout this stage reveals that it maintains a position on the *caudal* surface of the gastrula, not on its lower surface. This apparent discrepancy is explained by a tilting of the gastrula—as in the amphioxus—whereby the animal pole comes to face forwards and the vegetal pole is directed backwards (caudally). Thus the blastopore closes over the true vegetal pole, with the neural plate and notochordal plate lying on the new dorsal aspect of the gastrula.

The notochordal plate disengages itself from the adjacent cells in the roof of the archenteron and forms a solid cord below the neural plate. The latter develops into the neural tube in rather a different fashion from that of the amphioxus; the junctional region of neural plate and ectoderm rises up along each edge of the plate to form a *neural fold* with a *neural groove* between them. Eventually the folds meet over the groove; at this stage, the neural tissue separates from the ectoderm and forms a tube while the ectoderm from each fold fuses in the midline and covers the neural tube. Some cells in the crest of each neural fold go their separate way to aggregate lateral to the midline between the neural tube and ectoderm. At first as a continuous cord of cells on each side, these *neural crest cells* (Fig. 4.10) later aggregate into segmentally arranged masses giving rise to a variety of cells and tissues but particularly to the dorsal root ganglia.

Other features occurring during this phase are the disappearance of the original blastocoel, the increase in size of the archenteron as it becomes saddle-shaped astride the large endodermal mass, and finally the development of the mesoderm. The fate map shows the presumptive mesodermal cells lying in the lateral lip of the blastopore near the presumptive notochord. As the blastoporal edge grows downwards, the mesodermal cells turn inwards alongside the notochordal cells and take up a position in the plane between ectoderm and endoderm to proliferate there as a complete mesodermal layer. Further differentiation of the mesoderm provides segmentally arranged condensations, the *somites*, alongside the notochord and neural tube. (Notice the difference between the amphibian and amphioxus 'somites'; in amphioxus, the somite is the vesicle detached from the roof of the archenteron, its walls providing the mesodermal layer of the embryo: in the amphibian, the somite although it may later have a temporary cavity, is only a part of the mesodermal layer.) Non-segmented mesodermal condensations immediately lateral to the somites form the *intermediate mesoderm*. The remainder of the mesoderm—the *lateral plate*—which encircles the endodermal cell mass and separates it from ectoderm, splits into two layers, one applied to the endodermal cells and the other to the ectoderm, with the *coelom* between the layers. The mesoderm fails to penetrate between ectoderm and endoderm in those areas where the mouth and anus will develop.

Elongation of the frog gastrula to reach the primitive body form is super-imposed on the basic pattern; the lips of the blastopore, directed caudally, continue to proliferate and lay down ectodermal, neural, endodermal and mesodermal cells, but the bulk of the tissue in the caudal part of the frog embryo is mesodermal. The head of the embryo also arises mainly from proliferation of the mesoderm around the primitive mouth and from growth of the cephalic end (brain) of the neural tube. The cavity of the archenteron, originally small, is nevertheless a cavity surrounded by endodermal cells which remain as the lining of the alimentary canal after the yolk material within them has been used by the developing embryo. As proliferation slows down in its lips, the blastopore reduces to a short longitudinal slit, closes and finally disappears.

In the frog, therefore, the major adaptations of the fundamental pattern of development are closely associated with the increased store of yolk in the ooplasm.

THE BIRD (CHICK)
(Fig. 5.4)

It is hard to believe that the 'yolk' of a hen's egg is an ovum i.e. a 'single cell', enclosed in a cell membrane and surrounded by a vitelline membrane; its nucleus and cytoplasmic organelles are confined to one tiny spot immediately below the cell membrane. After fertilization and the production of two diploid nuclei, cleavage of the cytoplasm begins but only just; it is represented by a microscopic furrow between the first two cleavage nuclei and proceeds no further. When these two nuclei divide, a second furrow appears crossing the first at right angles. With every subsequent nuclear cleavage, another furrow is interposed between the nuclei involved until there is a growing network of tiny furrows with a nucleus in every mesh. Each nucleus and its surrounding cytoplasm is roofed and walled by cell membrane but there is no 'floor'—the contents are continuous with the underlying yolk. With continued nuclear cleavage, the area occupied by the nuclei and intervening furrows increases in diameter until it appears as a white spot on the surface of the yolk. In fact, a hen's egg, if fertilized, has normally developed this far at the time of laying. By this stage, the central 'cells' of the area have been able to partition themselves from the yolk and become complete cells. And so it goes on—peripherally, further nuclear division with partial cell formation and centrally progressive completion of cell formation. Meanwhile a pool of fluid develops under these central cells, raising them off the yolk and giving the area a translucent appearance—the *area pellucida*—in contrast to the opaque white of the periphery— the *area opaca*. The fluid forms the *sub-blastodermic vesicle* and the spreading patch of cells is the *blastoderm* or *blastodisc*. At first, there is only one layer of cells; later, some of the cells slip out from among their neighbours and create a new layer immediately underneath. The two layers, closely applied to one another, are named the *epiblast*—the upper—and the *hypoblast*—the lower. Both take part in the rapid expansion of the disc—a process which may be regarded as a bold attempt by the cells, to encircle the whole yolk. After all, the pattern of gastrulation inherited from amphibian ancestors required that the yolk be enclosed within the primitive body form of the embryo, but for the chick, this was obviously not feasible and the compromise was a further adaptation: only the area pellucida, and, in fact, only its central part develops into the embryo, allowing the cells of the area opaca to continue with their envelopment of the yolk.

In spite of the development of the blastodisc, there are still features comparable with those of amphioxus and amphibian. Remember that the upper part of the frog blastula is chiefly presumptive ectoderm with a crescent of cells on one side comprising presumptive notochord, neural tube and mesoderm. If we were to imagine this dome of cells flattening to form a disc, then the role of the epiblast in the embryonic area of the blastoderm of birds is clear—it, too, consists chiefly of the presumptive ectoderm with a crescent of presumptive notochord, neural tissue and mesoderm occupying one sector of the circumference (Fig. 5.4). The rest of the amphibian blastula (endoderm) is represented in the chick by the hypoblast. The basic features in amphibian gastrulation are proliferation and invagination of the presumptive notochordal and mesodermal cells and the same is true of the chick embryo. The main difference lies in the site of the invagination of these cells. There is no doubt about where the notochordal cells will enter the space between epiblast and hypoblast—it will be in the midline. But the mesodermal cells do not invaginate as in the amphibian. In the chick, they actually migrate medially and caudally from both sides to aggregate caudal to the position of the notochordal cells and create a midline thickening (the *primitive streak*) which grows rapidly in length as more and more presumptive mesodermal cells continue to aggregate in the midline. In this way the outline of the blastodisc changes from circular to pear-shaped. The anterior (cranial) end of the primitive streak is occupied by an aggregation, the *primitive (Hensen's) node or knot*, of notochordal cells while the rest of the streak consists of mesodermal cells. Thus the primitive streak, although a single, midline thickening, represents the dorsal and both lateral lips of the blastopore. With continued proliferation of these cells in the primitive streak, the next step is their migration into the plane between epiblast and hypoblast and an accompanying change in the contour of the primitive streak, i.e. it develops a groove along its whole length (Fig. 5.5) bounded on each side by a distinct fold or ridge. To achieve the dissemination of the cells to their definitive sites and particularly to allow the notochordal cells to be laid down in the midline of the disc as the precursor of the embryonic axis, the primitive streak appears to migrate caudally leaving the notochord (head process) to lie deep to the differentiating neural plate and neural tube and allowing the mesodermal cells to migrate laterally and cranially in the same plane as the notochord.

The mesoderm penetrates everywhere except in an area immediately in front of the tip of the notochord; there, the endoderm is particularly thickened, forming the *prochordal plate* and adhering firmly to the ectoderm—hence the exclusion of mesoderm. Later, these adherent layers, known as the *buccopharyngeal membrane*, break down to form the primitive mouth of the embryo. Caudal to the primitive streak, a similar feature, the *cloacal membrane* appears, pushed backwards by the growth of the disc, and breaking down eventually to form the anus. Alongside the notochord, *somites* and *intermediate mesoderm* like those of the frog take shape; and *lateral plate mesoderm* extends to the edge of the embryonic area, condensing against ectoderm and endoderm to leave on each side a slit-like *coelom* or channel, parallel to the embryonic axis.

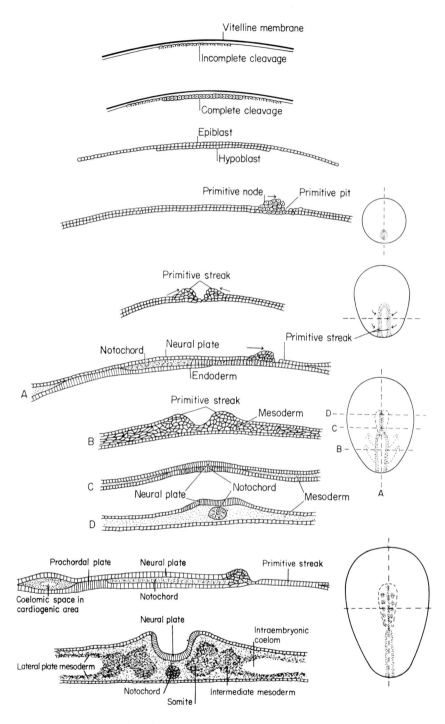

Fig. 5.4 Early development of chick embryo.

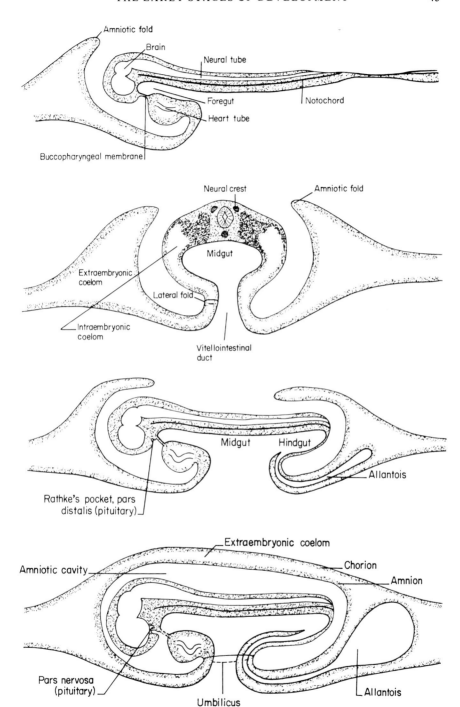

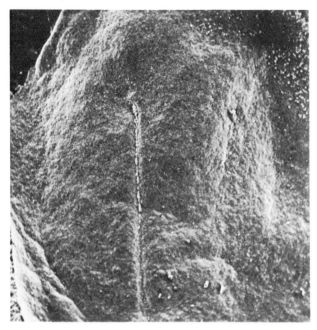

Fig. 5.5 Surface view of early chick embryonic disc showing primitive streak (by scanning electron microscope, 84x). By permission of Drs. Ruth Bellairs and Mary Bancroft: The Onset of Differentiation in the Epiblast of the Chick Blastoderm (S.E.M. and T.E.M.) in *Cell and Tissue Research*, **155**, 399–418 (1974).

These two spaces extend forwards in front of the prochordal plate to meet in the *cardiogenic area*; the whole coelom is therefore horseshoe-shaped.

Outside the embryonic area, an *extra-embryonic mesoderm* develops and proliferates between epiblast and hypoblast; it, too, separates into two laminae with an *extra-embryonic coelom* between them. Intra- and extra-embryonic mesoderm is in continuity around the embryonic area but their two coelomic cavities communicate only at the sides of the embryo.

Development of the primitive body form from these flat sheets of cells is an abrupt departure from the pattern inherited from the amphibian. It starts with the cephalic edge of the embryonic area turning under itself to form the *head fold*; this movement involves the part of the embryonic area in front of and including the buccopharyngeal membrane which therefore comes to face forwards and ventrally. The head fold drags the extra-embryonic part of the disc with it and also traps a portion of the sub-blastodermic vesicle; this pocket, lined with endoderm and bounded anteriorly by the buccopharyngeal membrane is the *foregut*. The cardiogenic area with its own part of the coelomic cavity now lies ventral to the foregut.

Along each side of the embryo and in continuity with the head fold, there develops a *lateral body fold* which involves the part of the embryonic area containing lateral plate mesoderm; and more of the sub-blastodermic vesicle becomes incorporated in the embryo to form, in this case, the *mid-gut*. Again, the extra-embryonic tissue is dragged in below the lateral folds. Much later, a *tail fold* takes shape, and creates a *hind gut*; the fold, itself, continuous of course with the lateral folds, involves only the cloacal membrane leaving the primitive streak on the dorsum of the embryo. These four folds meet ventrally but do not fuse; they leave an opening called the *umbilicus* in the ventral body wall; through it, the *vitello-intestinal duct*, lined with endoderm, leads from the gut to the sub-blastodermic cavity and to the yolk which is being slowly enclosed in a *yolk sac* by the extra-embryonic tissue.

The embryo now sits on top of the yolk sac, balanced precariously thereon and attached to it by extra-embryonic tissue. In fact, the embryo always turns over, usually on its left side, with the result that the umbilical opening is askew.

Although we set out to study the evolution of the early developmental stages in only four forms i.e. amphioxus, amphibian, bird and mammal, it is interesting to interpolate here a transitional step between amphibian and bird. In the former, the main features are complete cleavage of the ovum with epiboly of the ectodermal cells over the endodermal cells and hence the inclusion of the yolk (contained in cells) within the body of the embryo. In the chick, cleavage leads to the development of a blastodisc with embryonic and extra-embryonic areas. The embryonic area proceeds to the development of an embryo and, with the formation of head, lateral and tail folds, to the primitive body form; the extra-embryonic cells are left to enclose the yolk in an extra-embryonic yolk sac. However in the teleost fish where there is also the development of a blastodisc on 'top' of a non-cellular yolk mass, the blastodisc expands and envelops the yolk (i.e. by epiboly) with the development of a blastopore like that of the amphibian thus keeping the yolk within the primitive body form.

Early in the development of the chick, a fold of extra-embryonic tissue rises up in front of its head and creeps back over it like a hood. It consists of the extra-embryonic ectoderm with its adherent mesoderm and as it rises (Fig. 5.4) the extra-embryonic coelomic space increases in size under it. A similar fold, continuous with the head fold, arises along each side of the embryo, eventually reaching the caudal end where it joins with a similar fold coming up over the tail. When all these folds meet over the embryo, fusion occurs, the lower (inner layer) of each fold forming the *amnion* and enclosing the *amniotic cavity* around the embryo. The amnion is, of course, continuous with the ectoderm of the embryo at the umbilicus. The superficial (outer) layers of these folds fuse to form the *chorion*; its cavity is really the expanded extra-embryonic coelom. The chorion when fully developed encloses embryo, amniotic cavity, extra-embryonic coelom and yolk sac.

The *allantois* is another extra-embryonic membrane. It arises from the ventral aspect of the hind gut as a tiny diverticulum. As it increases in size, it protrudes through the umbilicus, alongside the vitello-intestinal duct, into the extra-embryonic space between amnion and chorion. Within a few days, it grows enormously in size, proceeds to 'take over' or occupy the whole extra-embryonic coelom and to fuse with the inner surface of the chorion. Although the functions of the extra-embryonic membranes described here are dealt with more fully in the next chapter (p.52) it is worth noting that they make their definitive appearance only at this stage in evolution viz. in birds and reptiles which lay their eggs in a 'dry' environment. Under these conditions the embryo would soon lose all its water content to the atmosphere and dry up without the protection of an 'artificial water jacket'.

THE MAMMAL (Fig. 5.6)

The progressive adaptation in the pattern of early development continues in the mammal, although its ovum, with very little yolk, is no larger than the amphioxus ovum. There are so many minor variations in the pattern among the mammals that it is easier to choose one, viz. the human, and follow its course of development, referring to those features which are significantly different in other types.

During cleavage, the human ovum reveals no sign of micromere or macromere formation. The *morula* stage, consisting of many small cells all of the same size and clustered inside a thick glassy membrane, the *zona pellucida*, is the first to show any differences in cell types; the peripheral layer of cells becomes the *trophoblast*, responsible from now onwards for the protection and nutrition of the embryo; the cells in the centre, known as the *inner cell mass*, will give rise to the embryo. Continued cell division and the accumulation of fluid inside the *trophoblastic shell* leads to the *blastocyst* in which the main feature is the adhesion of the entire inner cell mass to one part of the trophoblast; this mass then flattens to form a plate or *embryonic disc*, two cells thick, across a sector of the sphere, so that the space inside the trophoblast is divided into two unequal cavities; the smaller becomes the *amniotic cavity* overlying the embryonic disc, and the larger, foreshadowing the *yolk sac cavity*, lies below it; the two layers of the disc itself become *epiblast* and *hypoblast*.

Further development of the embryonic disc is so like that of the bird that it is difficult to distinguish the isolated embryonic disc of the bird from that of the mammal. The *amnion* continuous with and probably derived from the epiblast at the edge of the disc, is initially a layer of cells lining the inner surface of the trophoblast. In the same way, a layer of flattened cells continuous with the endoderm comes to form the wall of the *yolk sac*.

Extra-embryonic mesoderm also develops between the trophoblast and amnion and between the trophoblast and yolk sac wall. As the mesoderm thickens, it splits to produce an extra-embryonic coelom and two layers of mesoderm; one layer, clothing the amnion and yolk sac, is continuous at the edge of the disc with the intra-embryonic mesoderm; the other layer forms an inner lining for the

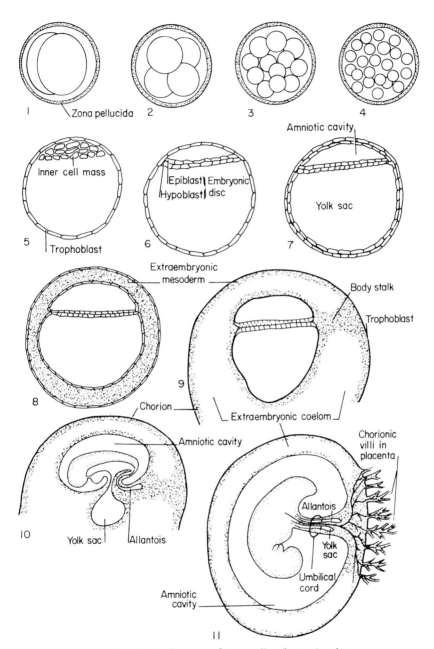

Fig. 5.6 Early development of mammalian (human) embryo.

trophoblast which, with this lining, is thereafter known as the chorion. But the extra-embryonic coelom does not completely separate the chorion from the embryonic disc, amnion and yolk sac; in the human, there remains a bridge of mesoderm, the *body stalk*, reaching from the edge of the disc to the chorion and responsible for carrying nourishment to the embryo.

The method of amnion development in the human, i.e. by *cavitation*, is not typical of the mammal. In the rabbit, for instance, the trophoblast overlying the embryonic disc disappears, and the disc, thus exposed, becomes intercalated in, and continuous with, the trophoblast layer. The surrounding trophoblastic tissue thereafter behaves as in the bird, i.e. it produces amniotic folds which grow over the embryo and fuse to give an amniotic cavity, an amnion, an extra-embryonic coelom and a chorion.

The *primitive body form* develops in the same way as in the chick, with the *vitello-intestinal duct* passing through the umbilicus from the gut to the yolk sac which in the human soon shrivels to a remnant closely applied to the body stalk. The yolk sac of mammals contains only a little fluid and no yolk; the sac develops because mammals inherited this feature from ancestors, e.g. reptiles, which required a store of yolk for their early development.

Although the *allantois* arises from the hind gut in the usual way, it, too, is poorly developed in the human, doing little more than burrowing into the substance of the body stalk. But the behaviour of the human allantois is not typical because, among mammals, there is great variation in the size and degree of development of both the yolk sac and the allantois—topics associated with the nutrition of the embryo and placenta formation.

FURTHER READING

Balinsky B.I. (1970) *An Introduction to Embryology* 3rd ed. Philadelphia & London: W.B. Saunders Company.

Brown A.L. (1970) *The African Clawed Toad.* London: Butterworths.

Ebert J.D. & Sussex I.M. (1970) *Interacting Systems in Development* 2nd ed. New York & London: Holt Rinehart & Winston, Inc.

Freeman W.H. & Bracegirdle B. (1967) *An Atlas of Embryology* 2nd ed. London: Heinemann Educational Books Ltd.

Hamilton H.L. (1952) *Lillie's Development of the Chick* 3rd ed. New York: Henry Holt & Company.

Huettner A.F. (1949) *Fundamentals of Comparative Embryology of the Vertebrates* 2nd ed. New York: The Macmillan Company.

Nelsen O.E. (1953) *Comparative Embryology of the Vertebrates.* London: Constable.

Romanoff A.L. (1960) *The Avian Embryo.* New York: The Macmillan Company.

Sussman M. (1964) *Growth and Development* 2nd ed. Englewood Cliffs, N.J.: Prentice-Hall, Inc.

Torrey T.W. (1967) *Morphogenesis of the Vertebrates* 2nd ed. New York & London: John Wiley & Sons, Inc.

6 PROTECTION AND NUTRITION OF THE EMBRYO

There are two contrasting methods of ensuring the survival of the species. The more primitive is the provision of no protection for the ovum or embryo, with, as a compensating device, the production of so many embryos that some are certain to survive. The second method is to produce only a few ova but to provide security for each embryo in the form of protective membranes or safe passage by keeping the embryo within the mother to an advanced stage of development.

Primary and *secondary egg membranes*, provided by the ovary or ovum and variously known as the vitelline membrane, fertilization membrane, or the zona pellucida, do not afford much protection. The more useful *tertiary membranes* are secreted by the oviducts. In amphibia, for example, the envelope of mucin or albumen is a conspicuous feature; it swells after the egg is deposited in water and holds water for a long time if threatened with desiccation. In reptiles and birds, the tertiary membranes are more vital because development normally occurs in a dry atmosphere. Thus, after fertilization, the ovum acquires several coats of albumen and, near the lower end of the oviduct, two distinct shell membranes are laid down around the albumen. In the reptile the outer one becomes thick and leathery but in birds calcareous crystals are deposited on it to form a brittle porous shell. In some mammals, tertiary membranes may be present but are insignificant.

The *extra-embryonic membranes*, i.e. amnion, chorion and allantois are derived from tissues of the zygote in birds, reptiles and mammals.

Every embryo needs food, oxygen and a method of excreting carbon dioxide and other waste products. The early amphioxus embryo is so small that all its cells can absorb food and oxygen from the surrounding water and also diffuse waste products (including carbon dioxide) into it; at the primitive body form stage these embryos can obtain food from water passing through the gut and excrete by the same route. The amphibian embryo can afford to remain within its albuminous coat to a more advanced stage because it has enough food in its endodermal cells to last well into the tadpole stage: and apparently enough oxygen and carbon dioxide exchange is possible by diffusion through the albuminous coat.

The bird and reptile, developing on land in their *cleidoic* (enclosed) egg, inaugurate a protective and nutritive system (Fig. 6.1) which is inherited and

51

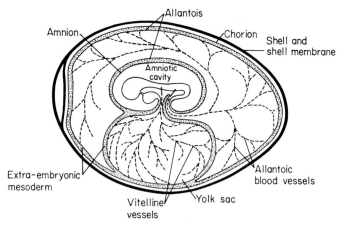

Fig. 6.1 Extra-embryonic membranes of chick.

modified by the mammals. After the development of its amnion and chorion (page 47) the chick embryo, surrounded by amniotic fluid, is in no danger of dehydration; it can also move freely, buffered from trauma, temperature variations and even from toxins. The extra-embryonic coelom is an additional water jacket bounded by the chorion which separates it from the albumen. After the albumen (and the fluid it holds) are absorbed into the extra-embryonic coelom and the amniotic cavity, the chorion lies close to the shell membrane, and surrounds everything within the egg.

Vitelline (yolk) *arteries* and *veins* develop in the mesoderm of the yolk sac wall and because they are linked to the blood circulation of the embryo, they carry the food material from the yolk sac into the embryo. We seldom stop to think how precise the stores of yolk are in the ovum of a bird or reptile—enough to last until hatching, when the remains are drawn into the abdominal cavity to tide the animal over the next 24 hours. Even more remarkable is the composition of the yolk; protein, carbohydrate, fat and other constituents are correctly proportioned to suit that particular bird or reptile and possibly quite unsuited for the development of any other embryo.

Respiration, a problem for the cleidoic egg, is solved by the development of the *allantois* (page 47). From the beginning, the allantois has a profuse network of blood vessels in its covering mesoderm and as it balloons out to fill the extra-embryonic coelom, it applies itself to the inner surface of the chorion to form the *chorioallantois*. The blood vessels of the allantois, linked to the circulation of the embryo are thus brought close up to the chorion, shell membrane and shell. Since the shell membrane and shell are porous, these chorioallantoic vessels can exchange oxygen and carbon dioxide with the atmosphere and act as a 'lung' for the embryo. An intriguing example of how Nature conserves her resources is illustrated during the development of the bird embryo; thus, much of the calcium salts in the shell is mobilized by the embryo through its allantoic vessels and incorporated into its bones.

Nitrogenous waste products are withdrawn from the blood stream by the kidneys, carried to the cloaca and thence to amniotic cavity and allantois, but if they were excreted as urea or ammonia, which are soluble in tissue fluids, these substances would diffuse throughout the tissue fluids outside and inside the embryo probably reaching lethal concentrations by the end of incubation. In the cleidoic egg, however, the embryo excretes its nitrogenous waste as uric acid which is deposited as insoluble crystals within the allantois; as well as acting as a lung, the allantois therefore serves as an excretory reservoir for the fetus.

THE PLACENTA

The mammalian embryo inherits these methods of 'breathing', obtaining its food and excreting its waste products but it has to modify them because the mammalian zygote has no yolk and no opportunity of respiratory exchange with the atmosphere.

The chorion which consists of trophoblast and extra-embryonic mesoderm, surrounds embryo, amniotic cavity, yolk sac, allantois and extra-embryonic coelom: because of its position, it provides the means whereby embryonic and maternal tissues are associated to form the *placenta* (see below). We must deal first, however, with the modifications which mammals create inside the chorion. Many use the yolk sac and the mesoderm intervening between it and the chorion to act as a bridge carrying its embryonic blood vessels to and from the chorion— a *chorio-vitelline placenta* (Fig. 6.2): this may be a temporary or permanent measure. In rodents, however, the mesoderm between yolk sac and chorion fails

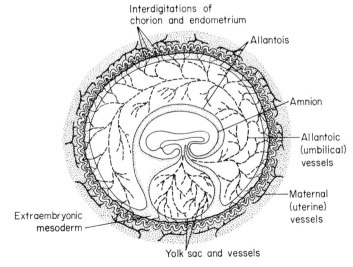

Fig. 6.2 Placenta (epitheliochorial, chorio-vitelline) and extra-embryonic membranes of pig embryo.

to develop, allowing the apposed yolk sac endoderm and trophoblast to
disappear: the remainder of the yolk sac then inverts ('turns inside out') to
bring the inside of the embryonic half of the sac against the uterine epithelium—
the *inverted yolk sac placenta* (Fig. 6.3). Many but not all animals supplement

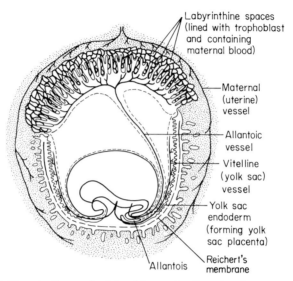

Labyrinthine spaces
(lined with trophoblast
and containing
maternal blood)

Maternal
(uterine)
vessel

Allantoic
vessel

Vitelline
(yolk sac)
vessel

Yolk sac
endoderm
(forming yolk
sac placenta)

Reichert's
membrane

Allantois

Fig. 6.3 Placenta (haemochorial and inverted yolk sac) and extra-embryonic
membranes of rat embryo.

or replace their yolk sac placenta by using the allantois (like the bird and
reptile) to take its blood vessels to the chorion and establish a *chorio-allantoic
placenta*. Probably the human modifies the pattern most; the yolk sac shrivels
up and the allantois fails to grow as far as the chorion (Fig. 6.4); hence the need
for a bridge or body stalk of mesoderm between embryo and chorion, along
which the allantoic vessels, on their own, continue to grow to reach the chorion.

Clearly the chorion plays the important role in nutritional and excretory
exchanges between fetus and mother; it reacts in different ways, however, on
meeting the endometrium. Some blastocysts seek a *central implantation* in the
uterus, i.e. each settles down in the lumen or in a crevice just off the lumen.
The behaviour of the trophoblast thereafter varies with the animal; in the pig, for
instance, *all* the trophoblast cells become applied to the epithelium, following
closely the folds and pockets all the way round the lumen, and the mesoderm
of the chorion is vascularized to the same extent (Fig. 6.2). Thus the tissues
intervening between the fetal and maternal blood, i.e. the *placental barrier*, are
the endothelium of the fetal blood vessels, the mesenchyme of the chorion, the
trophoblast, the epithelium and connective tissue of the uterus and the endothelium of the maternal blood vessels. Because all the chorion is used and because
the uterine epithelium is in contact with the trophoblast (chorion) this is known
as a *diffuse epitheliochorial placenta*.

In others, e.g. sheep and cattle, the chorion is not uniformly involved in

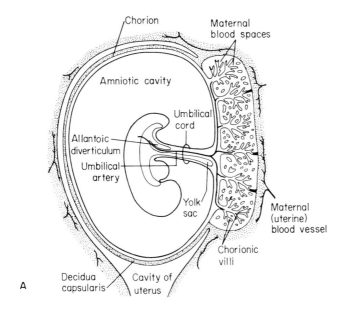

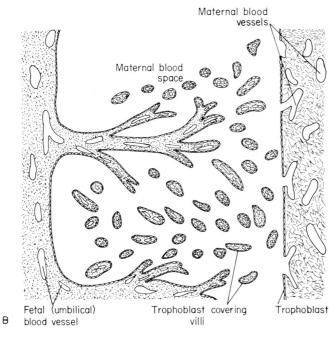

Fig. 6.4 A, placenta (haemochorial) and extra-embryonic membranes of human embryo, and B, a histological section through the human placenta showing the relationships of the maternal and fetal blood circulations.

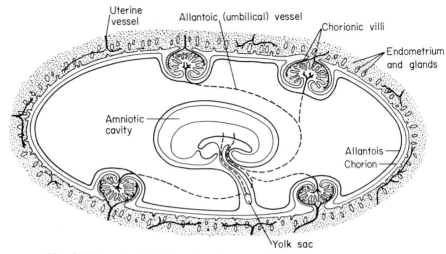

Fig. 6.5 Placenta (cotyledonary) and extra-embryonic membranes of sheep (or cow).

placental transfer (Fig. 6.5). Instead, there are, dotted over the chorion, numerous islands (or *cotyledons*) where the fetal vascular tissue is abundant, and the endometrium, lying opposite, is correspondingly thickened; this is the *cotyledonary* type of placenta and the apposed surfaces—epithelium and chorion—interdigitate deeply in the cotyledons. Most placentae show some kind of localization of the true placental area(s), e.g. the *zonary* or *annular* type (Fig. 6.6) (e.g. dog) and the *discoidal* (human) (Fig. 6.4).

More important however are the fine structural and functional variations in the placental barrier. In the cotyledonary type, for instance, the trophoblastic

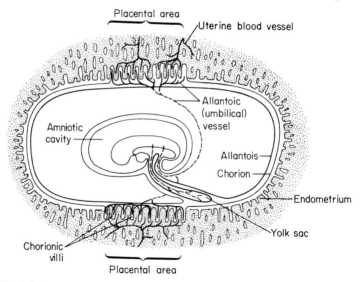

Fig. 6.6 Placenta (endotheliochorial) and extra-embryonic membranes of cat.

projections appear, in places, to erode the uterine epithelium and apply themselves to the connective tissue giving a *syndesmochorial* relationship; it is unlikely that this condition is sufficiently extensive in any placenta to warrant considering it as a distinct type. In the cat and dog, trophoblast invades still further to reach the endothelium of the maternal blood vessels—an *endotheliochorial* placenta. In the human and many other animals, the trophoblast destroys even the walls of the blood vessels allowing the blood to escape and form pools or lakes of blood around the chorionic villi or projections—a *haemochorial* placenta (Fig. 6.4). This extravasated blood is not stagnant and does not coagulate; as it escapes from the arteries, it is retrieved at the same rate by the veins. But the development of the human haemochorial placenta has special features; when the blastocyst enters the uterus, its trophoblast adheres to one area of endometrium; almost immediately, it starts to erode (and digest) the epithelium, the underlying connective tissue and finally the walls of the blood vessels. In this way, the whole blastocyst buries itself in the endometrium and becomes surrounded by blood—*interstitial implantation*. Meanwhile, the epithelium grows over to close the site of entry. There is then an intense proliferation of the trophoblast, but not uniformly; it grows like a sponge with an anastomosing meshwork of channels penetrating into its deeper layers from the periphery. The strands or trabeculae sometimes grow as branched villous processes with free ends (Fig. 6.4) but the final result is the same. The blood of course percolates through the spaces in the spongework or between the villi giving up food and oxygen and collecting soluble waste products. Finally, the trophoblastic strands are invaded by the extra-embryonic mesoderm and blood vessels to bring the embryonic circulation close to the maternal.

The chorionic sac with the embryo inside, however, is growing rapidly and, covered by a thin shell of endometrium, bulges into the lumen of the uterus (Fig. 6.4). Accordingly, the trophoblastic spongework over that part of the chorion which is covered by this shell of endometrium begins to regress; but it continues to grow where it is in contact with the underlying vascular endometrium, finally forming a discoidal area of placental tissue about 3 cms thick. Placentation and placental morphology are extremely complex subjects and the foregoing should be regarded only as an introduction.

The basic functions of the placenta are to nourish the fetus, to allow respiratory exchange and to act as a kidney. Exactly how it achieves all these is not quite clear. Nowadays, the view that placental transfer is achieved by diffusion and depends on the area of feto-maternal apposition and the thickness of the placental barrier is not popular because a great deal of evidence has accumulated in favour of active transportation of different ingredients—apparently with some degree of selection—from one circulation to the other. The role of the placenta as an immunological barrier is discussed on page 160 and its function as an endocrine organ on page 122.

We have referred to the condition of *oviparity*, i.e. the release of an ovum which is immediately independent of the parent, finding food for itself or surviving on its store of yolk; and also *viviparity*—retention of the ovum inside the uterus with complete dependence of the embryo and fetus on the mother for food and oxygen. But there is also the condition of *ovoviviparity*—i.e. the ovum, although it has large stores of yolk and uses them as food, is retained within the uterus, where it survives by means of oxygen and carbon dioxide exchange with the mother.

FURTHER READING

Assali N.S. (1968) *Biology of Gestation* Vol. 1. New York & London: Academic Press.

Barnes A.C. (1968) *Intra-uterine Development.* Philadelphia: Lea & Febiger.

Boyd J.D. & Hamilton W.J. (1970) *The Human Placenta.* Cambridge: W. Heffer & Sons Ltd.

Dawes G.S. (1968) *Foetal and Neonatal Physiology.* Chicago: Year Book Medical Publishers, Inc.

Hamilton W.J. & Mossman H.W. (1972) *Human Embryology* 4th ed. Cambridge: W. Heffer & Sons Ltd.

Nelsen O.E. (1953) *Comparative Embryology of the Vertebrates.* London: Constable.

Parkes A.S. (1952) *Marshall's Physiology of Reproduction* Vol. II 3rd ed. London: Longmans.

Romanoff A.L. (1960) *The Avian Embryo.* New York: The Macmillan Company.

7 ORGANOGENESIS

THE NERVOUS SYSTEM

The early neural tube (page 31 and Fig. 4.10), consisting first of tall columnar and later pseudostratified columnar epithelium, proceeds to form three distinct layers: the *ependymal layer*, nearest the lumen, is identified by the presence of many mitotic figures; later it shrinks to a simple epithelium. The *mantle layer*, which thickens slowly, contains all the true nerve cells of the neural tube with, of course, their quota of supporting tissue. In the adult, it becomes the *grey matter*. The outermost, the *marginal layer* is at first very thin, but as the *axons* (fibres) from the cells in the mantle layer push out and link up with neurones at other levels of the tube, they all lie in the marginal layer, and because of the myelin sheaths around the fibres, this layer becomes the *white* matter. The thin *roof* and *floor plates* of the tube are easily seen because the mantle layer does not encroach on them. The mantle layer, however, is subdivided into ventral or *basal* and dorsal or *alar laminae* (Fig. 4.10C), easily seen as separate swellings in transverse sections of the tube. In fact, these laminae are columns of cells running the length of the tube—hence the names given to their derivatives, the *ventral* and *dorsal grey columns* of the spinal cord.

There is also a functional segregation of the neurones within the nervous system. The nerve cells of the *neural crest* (Fig. 4.10B) bring impulses along their dendrites from all parts of the body; these are the true *sensory nerve cells* and their axons transmit the information by synapses to the neurones in the alar lamina. The neurones, there, transmit the impulses to the nerve cells in the basal lamina. From these, i.e. the true *motor (efferent) nerve cells* (Fig. 7.1), axons emerge to innervate muscles and glands.

Although often referred to as *afferent* the nerve cells of the alar lamina are 'middle-men'—*internuncial* or *intercalatory neurones* (Fig. 7.1). Their importance cannot be overestimated—they are responsible for inhibiting or strengthening, integrating and selecting all the impulses they receive from the sensory neurones before transmitting the results of their computations to the motor cells. The more complex the system of internuncial neurones the more complex the behaviour of the organism; its 'brain' consists almost entirely of alar lamina neurones.

The neural connections with the skin and skeletal muscles form the *somatic*

59

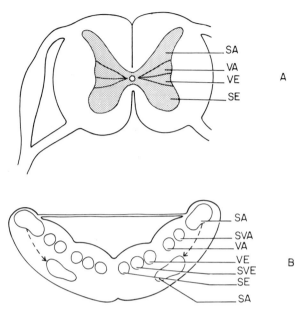

Fig. 7.1 Mantle layer (grey matter) of (A) spinal cord and (B) hindbrain, subdivided according to function of neurones. SA, somatic afferent; SE, somatic efferent; SVA, special visceral afferent; SVE, special visceral efferent; VA, visceral afferent; VE, visceral efferent.

innervation, those with the heart, lungs and intestine, the *visceral innervation*. For each, there are sensory, internuncial (afferent) and motor (efferent) neurones and in the spinal cord, separate columns of internuncial and motor neurones exist for both the somatic and the visceral innervations (Fig. 7.1). Furthermore, in the head region *special visceral* or *branchial columns* (both *afferent* (internuncial) and *efferent* (motor)) (Fig. 7.1) develop for the innervation of the visceral arches (Fig. 7.11).

Early in development, the front end of the neural tube shows three swellings or vesicles—*forebrain* (*prosencephalon*), *midbrain* (*mesencephalon*) and *hindbrain* (*rhombencephalon*) (Fig. 7.2), separated from one another by slight constrictions but the hindbrain gradually merges with the spinal cord. Next, a series of flexures appears (Fig. 7.2); two of these, the *midbrain* and *cervical flexures* are concave ventrally while the third, the *pontine flexure*, compensatory in nature, is concave dorsally and occurs in the hindbrain. Associated with the pontine flexure, there is a stretching of the roof plate, and dilatation of the lumen, with the two halves of the neural tube forming a flat floor instead of the side walls.

Early in evolution, the first step towards a greater awareness of the surroundings than is possible from ordinary sensory impulses from the skin was the development of a mechanism to inform the animal of its orientation and movement in the water. The *lateral line system* (page 80, Fig. 7.19), which became concentrated in the head region as the precursor of the *vestibular apparatus* in the *internal ear*, provided the information. These special sensory impulses and others received from skin, muscles and joints were collected in a conveniently sited 'integrating' and 'distributing' centre consisting of an extra mass of internuncial neurones near the cranial edge of the roof of the hindbrain (Fig. 7.2). The swellings caused by these cells form the *cerebellum*, which throughout evolution, retains control of movements concerned with balance and with the co-ordination of muscle action.

Probably the most important feature in the evolution of the brain is the successive appearance of higher and higher centres of integration, each developing cranial to the one before it. For instance, there are two integrating centres projecting on each side of the mid-line on the dorsum of the midbrain; together, in the human, they comprise the *corpora quadrigemina* (Fig. 7.2). The caudal pair of centres depend primarily on auditory impulses for their activities, the cranial pair on visual impulses and the relative sizes of these centres reflect the importance of the corresponding sensory organ in that animal.

The forebrain consisted initially of a central part and two cranial projections (vesicles) called the *olfactory lobes* receiving impulses from the nose. Later in evolution, the olfactory lobes moved backwards to project from the sides of the forebrain and became an integral part of it (Fig. 7.3). Together, the lobes and the cranial end of the forebrain form the *telencephalon*; the remaining central part is the *diencephalon*.

The diencephalon contains no true motor cells and therefore no representative of the basal lamina. Its mantle layer thickens to form the *thalamus* (Fig. 7.3), the original function of which is not clear but was probably an integrating centre for all sensory impulses below the diencephalon and perhaps for olfactory impulses as well. When the cerebral cortex evolves, the thalamus behaves as a relay centre for impulses passing to and from the cerebrum. The subsequent history of the original olfactory lobes (later known as telencephalic vesicles or *cerebral hemispheres*) consists of the development of the *corpus striatum* and the *cerebral cortex* (Fig. 7.3). Both are massive elaborations of the internuncial neurone system in the walls of the vesicles and the details of their genesis should be sought in more advanced texts.

A few general features are, however, worth noting. The cerebral hemisphere grows in an unusual fashion; on each side, it expands upwards (dorsally), then backwards (caudally) and finally sweeps round cranially and ventral to itself, in a C-shaped curve (Fig. 7.4).

Inversion of the original positions of *marginal* and *mantle layers* occurs by early migration of the mantle layer to the surface resulting in grey matter occupying a superficial position in the hemisphere and covering the white matter (Fig. 7.4). The enormous increase in the number of neurones in the cortex is achieved by the numerous foldings or convolutions of its surface, involving the whole thickness of the cell layer (Fig. 7.3).

The original lumen of the neural tube never disappears and in certain regions it expands to form ventricles (Figs. 7.1, 7.2, 7.3, 7.4): in each cerebral hemisphere, a *lateral* ventricle traces the path of the expanding hemisphere by becoming

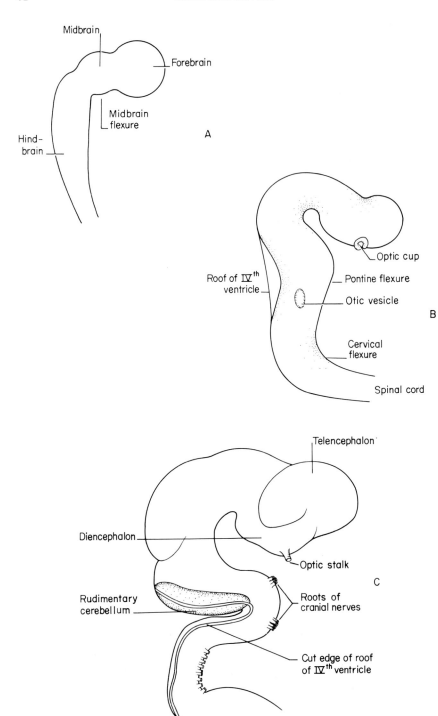

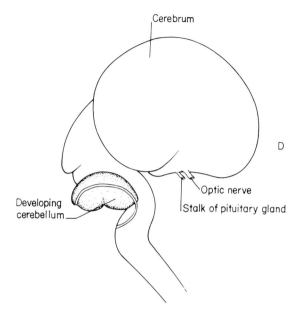

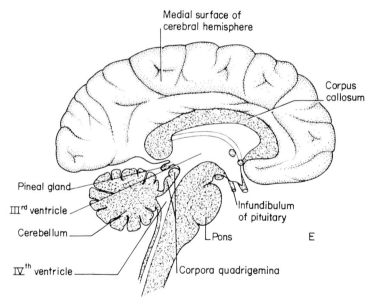

Fig. 7.2 Diagrams illustrating the development of the brain. A–D, lateral views; (E), medial view of midline sagittal section of adult brain. Stippled areas in (E) indicate cut surfaces.

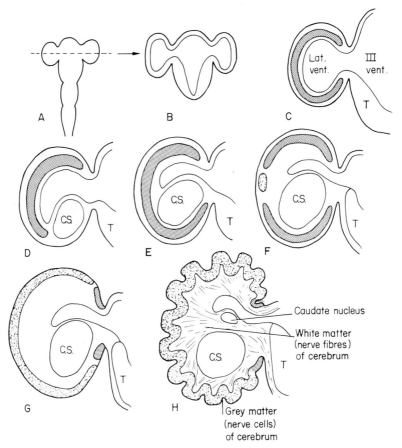

Fig. 7.3 Development of the cerebral hemisphere as seen in a series of transverse sections. CS, corpus striatum; T, thalamus.

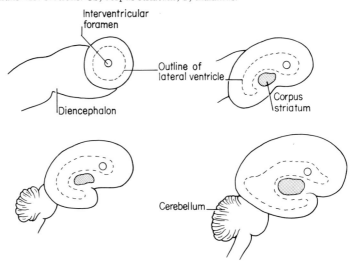

Fig. 7.4 Growth of cerebral hemisphere and lateral ventricle (side views).

C-shaped and communicates with the *third ventricle* sandwiched between the two thalami in the diencephalon; and in the hindbrain at the level of the pontine flexure the lumen distends as the diamond shaped *fourth ventricle*.

In this section, ontogeny has been blended with phylogeny to provide a better understanding of the pattern of development in the nervous system. Further details and the functional maturation of the brain are dealt with in Chapter 17.

THE EYE (Fig. 7.5)

A diverticulum protrudes forwards on each side from the floor of the diencephalon. Its free end becomes the *optic bulb* and the remainder the *optic stalk*, with an extension of the cavity of the diencephalon running through it into the bulb. Invagination of the distal free surface produces a double-walled *optic cup*; another, linear, invagination, continuous with the first, appears along the medial side of the bulb and continues into the adjacent part of the stalk. This long gap on the side of the cup and stalk, known as the *fetal* or *choroidal fissure*, soon closes leaving little or no trace, but the good reasons for its appearance and brief existence will be mentioned later. The space between the two walls of the optic cup and optic stalk disappears but the invaginated cavity of the optic cup continues into the stalk, coming to the surface where the fissure ends.

The three layers of the neural tube (page 31) are represented in the optic diverticulum. For example, the marginal layer can be traced, potentially, from the brain out along the stalk, on to the outside of the cup, over the lip to become a lining for the inner wall and back along the inside of the stalk. Nerve fibres are therefore confined to these positions. The outer wall of the optic cup, however, reduces to a single layer of melanin-containing cells, the *pigmented layer of the retina*, while the more important inner wall continues developing to become the *nervous layer of the retina*. The *rods* and *cones* of the retina are probably derived from the ependymal layer of the inner wall of the cup and become firmly apposed to the pigmented layer. The nerve cells of the retina develop in the original mantle layer and, when their nerve fibres emerge, they run in the marginal layer, i.e. nearest the inside of the cup.

The channel, leading from the inside of the cup into the optic stalk and remaining after the edges of the choroidal fissure have fused, provides an exit for nerve fibres and also creates the '*blind spot*' on the retina. As these fibres increase in number, they increase the thickness of the marginal layer lining the channel and occlude its lumen, and where this channel reaches the surface of the stalk at the proximal end of the fissure the nerve fibres also emerge and continue towards the brain on the surface of the stalk.

The ectoderm overlying the optic cup thickens to form the *lens placode*, which sinks below the surface to give a closed *lens vesicle*; the lips of the optic cup then trap the vesicle within the cup. The cells of the posterior wall of the

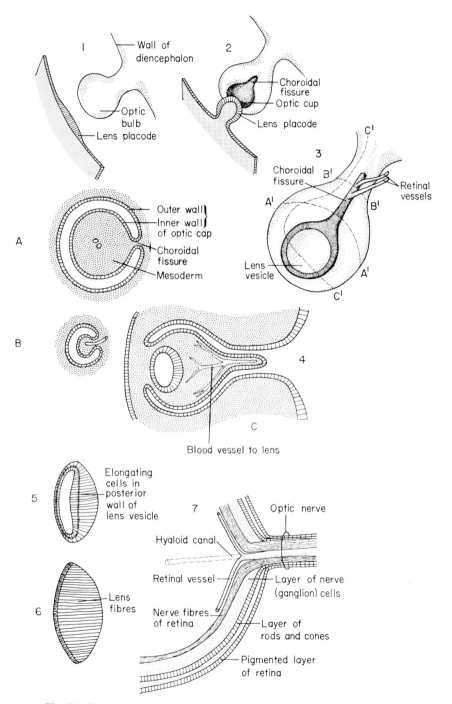

Fig. 7.5 Eye: development of retina and lens. (A), (B) and (C) represent sections through the eye at planes A^1A^1, B^1B^1, and C^1C^1 in diagram 3.

lens vesicle elongate, reducing and finally obliterating the lumen; these cells lose their nuclei and form the *lens fibres*. The front wall of the vesicle remains as a single layer of cuboidal cells, while the cells at the equator of the developing lens continue to multiply, supplying more lens fibres during the growing period.

The *choroid* and the *sclera* are differentiated from mesenchyme around the optic cup; in the choroid, the tissue remains loose and vascular, developing smooth muscle for the iris and ciliary body; in the sclera, it becomes a thick layer of dense connective tissue which is continued forwards as the *cornea* in front of the iris. Inside the eye, the aqueous and vitreous humor are also derived from mesenchyme. The blood vessels inside the eye are linked with those outside by vessels which take the same pathway as the nerves emerging from the retina, i.e. along the channel in the optic stalk. During growth of the lens, a vessel runs forwards from the blind spot, through the vitreous humor to nourish the lens, but when it atrophies, there remains in the vitreous a thin membranous tube called the *hyaloid canal.*

On reaching the diencephalon, half of the fibres in each optic nerve i.e. from the medial half of the retina, cross to the opposite side of the brain. This bundle of decussating fibres is dragged out of the brain substance until it forms the *optic chiasma*, a bridge, entirely free of the brain, between the two optic nerves. It is important to realize that the description and developmental process given above refers to the mammalian eye and that special features are present in other types. For instance, in the bird's eye a structure known as the pecten projects from the region of the optic disc into the vitreous humor. It comprises blood vessels and connective tissue enclosed in epithelium and probably develops in a comparable fashion to the hyaloid vessels which degenerate completely in the mammalian eye. The tapetum found in the eyes of nocturnal and some other animals as a reflecting layer in the choroid, consists either of glistening connective tissue fibres or of cells in the choroid containing guanine crystals.

Although the eye of the insect presents such a different gross morphology from that of the mammalian eye, close examination reveals that it develops from the same basic elements and is in fact an early evolutionary form. In the simple eye of the insect (Fig. 7.6), the cornea is a transparent part of the cuticle secreted by special (corneagenous) cells of the surface epithelium. As well as being transparent, this area is shaped to behave as a refractive element e.g. dome-shaped or biconvex. Deep to the cornea a further refractive element, probably representing the lens of the mammalian eye, is usually present; it may be a secretion of the corneagenous cells or may even be an aggregation of transformed corneagenous cells. The retina of the insect eye lies close to the cornea/lens and its cells are elongated at right angles to the surface. The light sensitive surfaces of adjacent cells are often apposed to one another forming the rhabdom which lies parallel to the long axis of the cell.

Even more primitively, the photoreceptors in some invertebrates are single cells differentiated from the surface epithelium: when, however, these cells retreat from the surface as they do in the more advanced forms, they must have the transparent material overlying them.

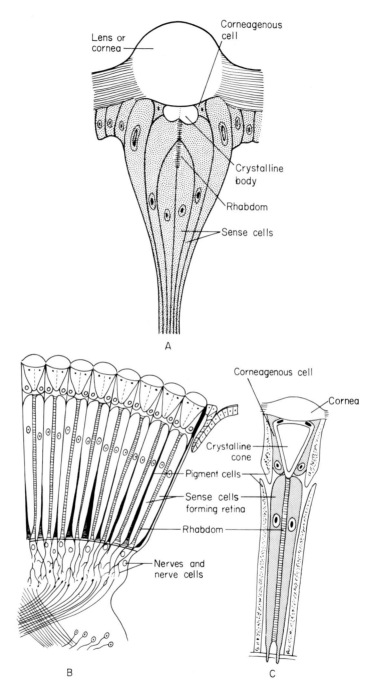

Fig. 7.6 (A), the simple eye of the insect and (B), the compound insect eye with (C), an individual ommatidium, for comparison with the structure of a vertebrate eye. From Ross, H.H., *A Textbook of Entomology*, 3rd edition. John Wiley & Sons, Inc. New York, London.

PITUITARY GLAND (HYPOPHYSIS CEREBRI) (Fig. 7.7)

The little hollow between the growing forebrain and the heart is the primitive mouth or stomodeum; from its roof, immediately in front of the bucco-pharyngeal membrane, a small ectodermal growth, *Rathke's pouch* burrows up through the mesoderm towards the brain. Its connection with the mouth

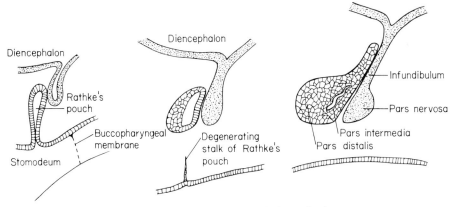

Fig. 7.7 Development of the pituitary gland.

disappears leaving a closed vesicle; the anterior wall of the vesicle thickens, its cells forming the *pars anterior* (*distalis*) and obliterating the cavity. The posterior wall also thickens slightly to give the *pars intermedia.* Meanwhile a smaller diverticulum from the diencephalon grows down to apply itself to the posterior wall of Rathke's pouch. It, too, loses its lumen and forms the *pars nervosa* of the pituitary while its connection with the brain remains as the *infundibulum.* The parenchymal cells of the pituitary form small clusters or cords, liberally supplied with blood capillaries. As might be expected, the pituitary gland varies in its detailed form among the vertebrates but the essential components and developmental features can be easily recognized in most cases.

The Pineal Gland (Fig. 7.2) begins as a diverticulum from the caudal end of the diencephalic roof; as its walls thicken, the lumen disappears to give a solid organ. The apparently simple structure and obscure function of this gland in mammals probably belies its intriguing evolutionary history during which it had a rudimentary eye-like structure and a light-perceiving function.

THE FACE AND NECK

The primitive mouth or *stomodeum* lies between the bulging forebrain above and the heart below (Fig. 7.8); the buccopharyngeal membrane, until it breaks down, forms the floor or inner wall of the hollow. On the side of the head, the primordium of the eye projects from the brain and in front of each eye a

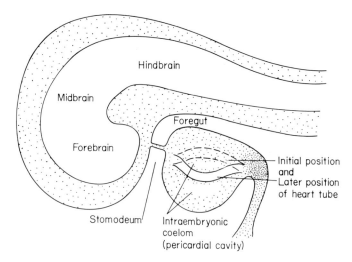

Fig. 7.8 Mid-line (sagittal) section through cranial end of early embryo.

thickened area of ectoderm, the *olfactory placode*, overlies the forebrain (Fig. 7.9). These are the landmarks to watch while the other elements of the face differentiate from the underlying mesoderm. In particular, the mesoderm around the olfactory placodes, between them, lateral to them and above them, proliferates to form an M-shaped swelling called the *fronto-nasal process* (Fig. 7.9); the middle 'leg' is the *median nasal process*, the two lateral 'legs', the *lateral nasal processes*, separating the nasal placode from the eye on each side.

Between the buccopharyngeal membrane and the heart the mesoderm raises a U-shaped ridge or collar with its free (dorsal) ends fading out on the sides of the head. This, the *first* or *mandibular arch*, is the first of a series of *six visceral* (branchial) *arches* which not only effectively increase the distance between mouth and heart but also the length of the foregut (Fig. 7.9). An outer *ectodermal groove* and an inner *endodermal pouch* delineate each arch from the next (Fig. 7.16). Since the first arch is so intimately associated with the development of the face we must take note now of the internal features of these arches. On each side, a blood vessel or *aortic arch* runs from the midline to the dorsal end of each arch (Fig. 7.10); dorsally, one or two *nerves* enter to supply the tissues (Fig. 7.11) while a U-shaped bar of *cartilage* acts as a 'skeletal support'. In the face, the mandibular arch has a spur-like swelling (the *maxillary process*), projecting forwards from its dorsal end below the eye (Fig. 7.9). These, in fact, complete the basic parts of the face.

The behaviour of the frontonasal process (Fig. 7.9) depends primarily on the olfactory placodes which refuse to be dislodged or raised from their position close to the forebrain; they remain as the *olfactory epithelium* in the roof of the nasal cavity. The three-pronged frontonasal process grows forwards and downwards like a double hood or cowl over the olfactory placodes leaving a channel which runs up on either side of the median nasal process towards the olfactory

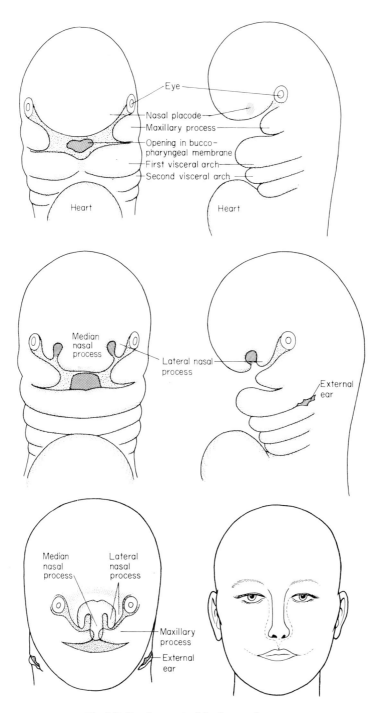

Fig. 7.9 Development of the human face.

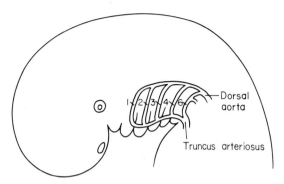

Fig. 7.10 The aortic arches on the left side of the embryo.

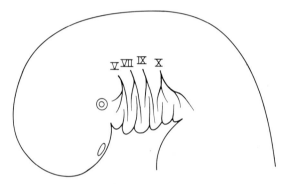

Fig. 7.11 The nerves supplying the visceral arches: V – Trigeminal, VII – Facial, IX – Glossopharyngeal, X – Vagus.

placode. The surface openings are the future nostrils, still incomplete inferiorly; the median nasal process, the future *nasal septum*, retains for a time (Figs. 7.12, 7.13), a free lower edge while the two lateral nasal processes will form the sides of the nose in the human; if the frontonasal process projects boldly it produces the snout or muzzle of the lower mammal, with an equivalent degree of growth in the rest of the face.

As the mandibular arch grows out in front of and below the buccopharyngeal membrane, it forms a *floor* for the adult *mouth cavity*. At the same time (Fig. 7.9), the maxillary process grows forwards and medially as an arch on each side below the eye to produce the side wall of the mouth or *cheek*; fusing with the side of the lateral nasal process and then slipping below its free lower edge, it continues to grow medially below the nostril to meet and fuse with the median nasal process and the opposite maxillary process; so form *the upper lip* and *upper dental arch* (Fig. 7.13). To separate the mouth and nasal cavities a horizontal shelf, the *palatal fold*, projects from the inner aspect of each maxillary process to fuse in the midline with its opposite number and with the free edge of the median nasal process, thus forming the palate (Fig. 7.12).

Variations in facial features, whether gross, like the differences between lower vertebrates and man, or minor, e.g. the familial human traits, are the result of differences in the amount of growth in the elements of the face. One striking example of gross variation occurs in the elephant where the region around the nostrils and upper lip undergoes unusual growth to form the trunk.

The foregoing description is based on the concept of a series of separate projections fusing with one another along embryonic suture lines. It has the advantages of being simple and of providing an understanding of normal development and of the common congenital defects occuring in this region. A more accurate representation of facial development visualizes the formation and fusion, of sub-ectodermal masses of mesoderm representing the nasal and maxillary processes in the conventional description. But further explanation and details should be sought in more advanced texts.

All the facial bones are formed in membrane; this is true of the *mandible* in spite of the cartilaginous bar (*Meckel's cartilage*) in the mandibular arch; the

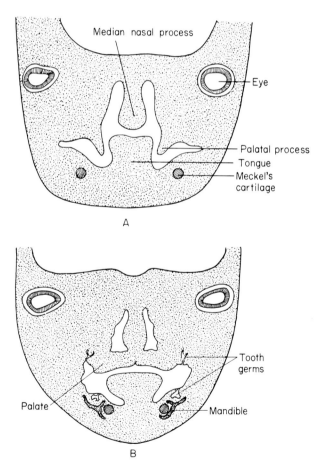

Fig. 7.12 Sections through the head to show positions of palatal folds, (A) before fusion and (B) after fusion with nasal septum.

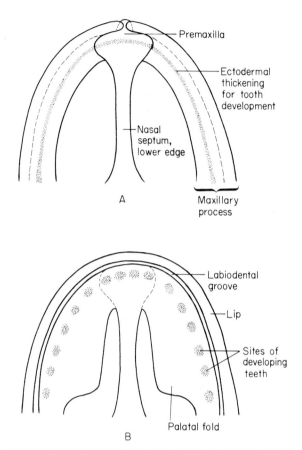

Fig. 7.13 Lower surface of upper dental arch, (A) before and (B) after development of the palatal folds.

bone develops superficial to and separate from the cartilage but encircles it later (Fig. 7.12). At its dorsal end, however, where the mandible lies near the skull, superficial to Meckel's cartilage, a nodule of cartilage develops, from which all growth in the length of the mandible occurs by means of endochondral ossification. This nodule behaves like an epiphyseal plate but it differs in that it persists into adult life as a dormant cartilaginous cap at the head (condyle) of the mandible. The coronoid process and prominent angle (Fig. 7.14) which give the bone its irregular shape are merely thickened projections where the muscles of mastication are attached, and the raised platform carrying the teeth (Fig. 7.14) is developed for holding the teeth and when these are removed, the platform is resorbed. Such adaptations in the shape of a bone are examples of what occurs regularly, if less spectacularly, throughout the skeleton during development.

The nerves to the visceral arches are unusual; they run towards the grooves or clefts between the arches, each dividing into *pretrematic* (precleft) and *post-trematic* (postcleft) divisions (Fig. 7.11); each arch therefore receives a

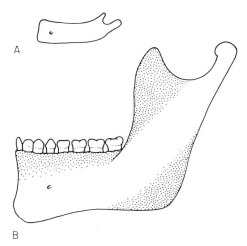

Fig. 7.14 The human mandible, (A) in the newborn and (B) in the adult. The shaded areas indicate bone developed in response to muscle attachments or to the presence of teeth.

contribution from two nerves. In the face region, the mouth is regarded as a cleft and the Vth cranial nerve sends its pretrematic branch (maxillary nerve) into the maxillary process; its post-trematic branch (mandibular nerve) enters the first arch to supply the muscles of mastication and ordinary sensation. The next (VIIth) nerve, straddling the cleft between first and second arches, sends its pretrematic branch (chorda tympani) into the first arch to supply glands and the sensation of taste.

The opposing edge of the mandibular arch and of the composite arch formed by maxillary processes and the median nasal process undergo special development (Fig. 7.13). In the mammal, an elongated thickening of ectoderm appears along each edge and burrows into the underlying mesoderm; this epithelial invasion eventually scoops out a ditch or channel known as the *labio-dental groove*; the outer wall forms the lip and the inner wall becomes the dental arch. Another ectodermal thickening appears on the crest of each dental arch but it breaks up into a series of epithelial clusters, each of which burrows into the mesoderm to form a *tooth bud*, a double layered cup inverted over a tiny mass of mesoderm (Fig. 7.15). The inner layer of epithelial cells (the *ameloblasts*) manufactures the *enamel* of the tooth while the mesodermal cells (*odontoblasts*) inside the cup produce the *dentine*.

From the thread of ectodermal cells initially connecting this tooth germ with the surface, a small bud grows out as the germ of the *permanent* tooth which lies dormant until after the first (*deciduous*) tooth has erupted.

Many variations occur in the pattern of development around the free edge of the mouth in different creatures, but the one consistent feature is the thickening of epithelium which can and does give rise to a variety of 'teeth'—from simple keratinized scales through primitive types of teeth to that of the primate with

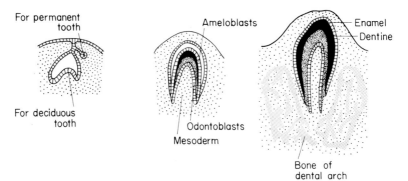

Fig. 7.15 Tooth development.

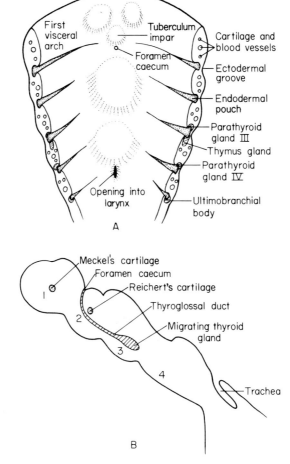

Fig. 7.16 Development of tongue, floor of mouth and pharynx: (A), (C) and (E) dorsal views of floor of pharynx; (B), (D) and (F) sagittal sections of floor of mouth and pharynx.

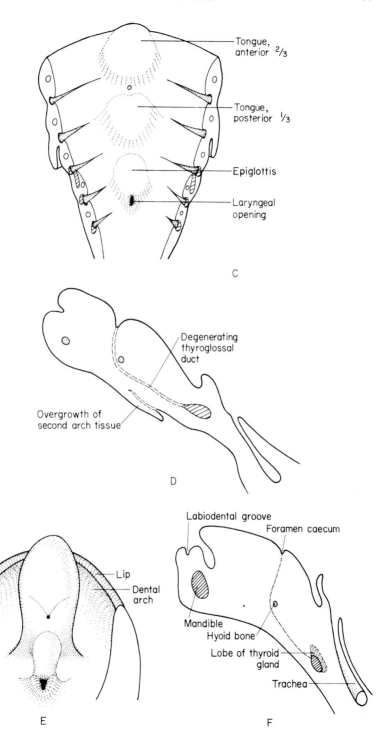

Tongue, anterior 2/3

Tongue, posterior 1/3

Epiglottis

Laryngeal opening

C

Degenerating thyroglossal duct

Overgrowth of second arch tissue

D

Lip

Dental arch

E

Labiodental groove

Foramen caecum

Mandible

Hyoid bone

Lobe of thyroid gland

Trachea

F

both ectodermal and mesodermal elements; an interesting example of adaptation are the large keratin 'scales'–*the beak*–projecting from the 'lips' of the bird.

Under the endoderm of the first arch two swellings close to the midline and another in the midline (tuberculum impar) arise, coalesce and produce the anterior two-thirds of the tongue (Fig. 7.16). The posterior one-third derives from the third arch which flows over the second arch to meet the anterior part; a V-shaped groove pointing caudally on the surface of the adult tongue marks the junction of the two parts. Thus the nerves of the anterior two-thirds of the tongue are the post-trematic branch of the trigeminal (Vth) for ordinary sensation and the pretrematic of the VIIth nerve for taste: the IXth (glossopharyngeal) nerve supplies the rest; the *musculature*, the main mass of the tongue, migrates from the *occipital myotomes* and brings its own nerve supply, the hypoglossal nerve.

The visceral arches

Our early ancestors used the 3rd–6th arches as true *gill arches*, with the intervening ectodermal and endodermal grooves meeting to give a series of *gill clefts*. Thin-walled capillaries sprout from the aortic arches and project, finger-like, from the edges of the clefts into the stream of water coming out of the pharynx (Fig. 7.17). The blood can thus exchange CO_2 and O_2 with the water. In fish, the opening of the first cleft is reduced to a small perforation called the *spiracle* which phylogenetically is a precursor of the middle/external ear. Since

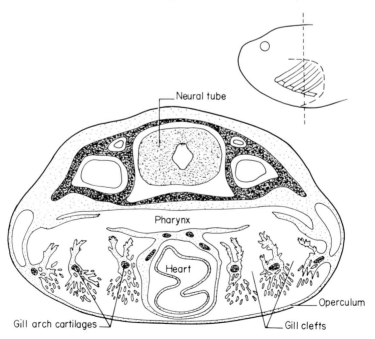

Fig. 7.17 Transverse section through gill region of a tadpole.

the gill clefts with their delicate vascular fringes are extremely vulnerable, the second arch grows backwards over them as a protective cover or *operculum*.

Terrestrial animals could not jettison this plan of development but they do *adapt* it during embryonic life. Gill arches with intervening grooves are formed but actual clefts do not appear. On the outside, the operculum, instead of forming a cover with a space between itself and the other arches, fuses with them as it grows backwards, smoothing and obliterating the ridged pattern.

We need not describe the fate of each separate arch; their cartilages give rise to the hyoid bone, the cartilages of the larynx and perhaps the upper rings of the trachea; the muscles of the pharynx and larynx as well as some of the more superficial muscles of the neck, e.g. sternomastoid, are recognized by their nerve supply as coming from the arches. The aortic arches are also adapted but they will be described with the vascular system.

We must look more closely at the *endodermal* or *pharyngeal pouches* in the later evolutionary forms (Fig. 7.16). In the midline of the first groove (between the two parts of the developing tongue), a pocket of endodermal cells burrows into the mesoderm; retaining their connection with the surface, the cells extend through the mesoderm of the second and succeeding arches until, below the level of the developing larynx, this cord of cells bifurcates to form two large masses, the primordia of the lobes of the *thyroid gland*. Each lobe consists of tiny cell clusters and each cluster develops a space or *thyroid vesicle* in its centre to store the secretion of its cells (Fig. 4.2). The original cord of cells, the *thyroglossal duct*, reaching from the tongue to the adult gland, normally disappears, but its origin from endoderm is marked by a pit, the *foramen caecum*, at the apex of the V-shaped groove on the back of the tongue. The adjacent portion of the first endodermal pouch and the corresponding part of the second become obliterated by the tongue primordia but the dorsal end of the first groove, associated with the spiracle in the fish, is involved in the development of the middle ear in higher vertebrates.

From the third endodermal pouch an endodermal ingrowth becomes the *thymus gland*, which burrows downwards in the neck—even into the thorax in some animals. It is still not clear, however, whether the lymphocytes of this gland arise from the endodermal cells or from the mesenchyme so intimately associated with them.

Close to, even confluent with, each thymic growth, another endodermal evagination gives rise to a *parathyroid gland*. The little nodule of cells loses its connections with the endodermal pouch, acquires the typical endocrine arrangement of its cells, and migrates caudally to a position dorsal to the thyroid gland.

From the fourth endodermal pouch, another pair of *parathyroid glands* develop and migrate in the same direction.

It is often difficult to distinguish the last two pouches but, reckoned as derivatives of the fifth, two further endocrine glands, the *ultimobranchial bodies* also develop. In birds, these remain as separate organs but in mammals, they become incorporated into the thyroid as the scattered parafollicular cells.

Functionally, the separate glands and the incorporated ultimobranchial cells are similar—they produce calcitonin, a hormone with opposite effects to those of the parathyroid hormone.

THE EAR (Fig. 7.18)

Very early in development a thickening of surface ectoderm (the *otic placode*) appears over the region of the hindbrain. It soon becomes a closed egg-shaped sac, the *otic vesicle*, which sinks into the underlying mesoderm to form the *membranous labyrinth* of the *internal ear*; first, it constricts to give two main parts, the *saccule* and *utricle*, joined by a narrow *ductus reuniens*. From the utricle three flattened leaf-like projections finally become the *semi-circular ducts* and from the saccule a single diverticulum protrudes, curling and spiralling to form the *cochlear duct*. The *otic capsule*, a condensation of the surrounding mesoderm, develops into the bony labyrinth which follows the contours of the membranous labyrinth but is separated from it by a narrow fluid-filled space. Overlying the internal ear, the *ectodermal* groove between the dorsal ends of the first and second visceral arches deepens to form the external auditory canal; the corresponding *endodermal* pouch balloons out as the middle ear cavity, retaining communication with the pharynx by a hollow stalk, the auditory (Eustachian) tube; the membrane separating the ectodermal groove from the endodermal pouch persists as the tympanic membrane. As the middle ear cavity grows, it encounters the ends of three visceral arch *cartilages*, viz. *Meckel's*, *pterygomandibular* (pterygoquadrate) and *Reichert's*, lying close together and in that order from lateral to medial. These cartilaginous tips become isolated from the remainder of their shafts, and invaginated into the middle ear. The lateral nodule forms the *malleus*, the middle one the *incus*, and the medial the *stapes*.

The pretrematic branch of the VIIth (facial) nerve on its way to the mandibular arch becomes pushed aside by the expanding middle ear cavity and finds itself embedded in the tympanic membrane just deep to the endoderm— hence the name *chorda tympani*.

Around the edge of the external meatus, a series of mesodermal tubercles grow out from the adjacent arches, to form the auricle or *pinna*. Only the tragus, the part in front of the external meatus, comes from the 1st arch, the remainder arising from the second.

The evolution of the ear explains much of its complicated development. In spite of its obvious preoccupation with hearing, the ear was originally a sensory organ, the *lateral line system*, associated with balance and orientation (Fig. 7.19). At first as a row of special cells in the epidermis of the fish, responding to the currents in the surrounding water, it later formed a closed tube; the sensation of changing direction and movement then depended on the inertia of the fluid within it. This single tube is eventually restricted to the cranial end of the animal, acquiring transverse and oblique extensions and loops around the head, thereby detecting all movements of the head. In some parts of the

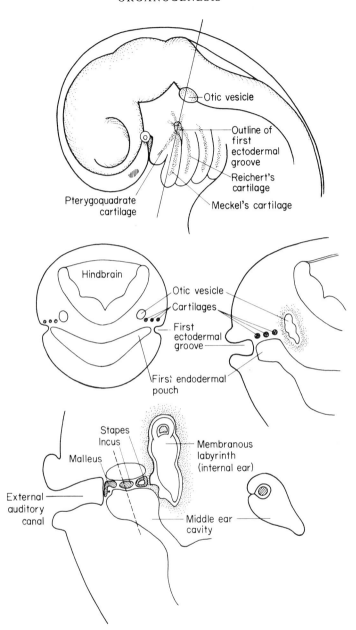

Fig. 7.18 The development of the ear. From Hamilton, W.J. and Mossman, H.W., *Human Embryology*, 4th edition. Wm. Heffer & Sons Ltd., Cambridge.

system, *otoliths* appear in the lumen (presumably secreted by the lining epithelium) and since they can respond to gravity their positions within the tubes can indicate the orientation of the head in space. The first reliable evidence of any auditory apparatus came with the loss of the spiracular cleft

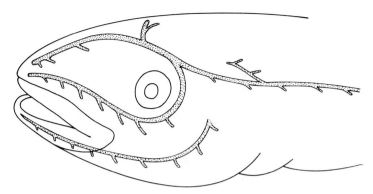

Fig. 7.19 The distribution and arrangement of the lateral line system in a fish (left side).

i.e. with retention of the membrane between the first ectodermal groove and first endodermal pouch. Part of Reichert's cartilage forms a rod with its lateral end against the tympanic membrane and its medial end touching the membranous labyrinth; even at this stage, the cartilaginous rod (*columella*) is invaginating the endodermal pouch. The appearance of a blunt diverticulum, the *lagena*, from the membranous labyrinth heralds the development of a cochlear duct. Progressively, thereafter, there are improvements in the mechanism of transmitting vibrations from the membrane, all designed to make it more efficient as well as less vulnerable.

THE CARDIOVASCULAR SYSTEM

All blood vessels and blood develop from mesenchyme, appearing first as tiny *blood islands*; these take shape when clusters of mesenchymal cells assemble to form closed vesicles containing rounded mesenchymal (primitive blood) cells. As the number of islands increases, they coalesce in an anastomosing meshwork of channels which provide the adult capillaries by persisting in that form, but the larger blood vessels also arise from them by dilatation and by thickening of their walls.

Heart

When the head fold of the embryonic disc swings into position ventral to the foregut, it carries the cranial loop of the *intra-embryonic coelom* with it to lie just cranial to the umbilicus but separated from it by that mass of mesoderm called the *transverse septum* (Fig. 7.8) (see p.70). In the mesoderm dorsal to this intra-embryonic coelomic loop and ventral to the foregut, two blood channels, lying close to the midline, fuse to form a single *heart tube*. Irregular twitchings of its wall start very early and develop into regular peristaltic waves travelling from behind forwards to provide an organised blood flow. At this stage, the wall of the heart is a very loose meshwork of mesenchyme with no sign of the muscle tissue which develops later. The heart tube then invaginates the roof of the coelomic space, carrying with it a sheet of mesoderm which soon disappears, and the heart, although anchored at each end, is then free to increase in length, coil or change its shape within the pericardial cavity. Even at this stage, certain features can be identified (Fig. 7.20): a sinus venosus on each side emerges from the transverse septum and empties into a common dilated part, the atrium; a permanent constriction, the *atrio-ventricular canal*, which will later contain the atrio-ventricular valve, separates the atrium from a second dilatation, the *ventricle*. Beyond the ventricle, the heart tube consists of first

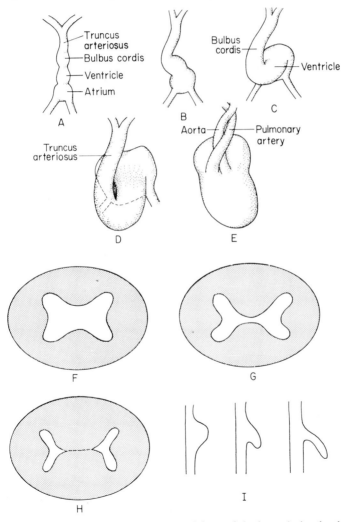

Fig. 7.20 (A)–(E), changes in the external form of the heart during development; (F)–(H), septation of the atrioventricular canal by anterior and posterior endocardial cushions; (I)–development of valve cusp from endocardial cushion.

the *bulbus cordis* and then the *truncus arteriosus* which divides to send a branch round each side of the foregut. This single tube, beating regularly and pushing the blood cranially, is enough for creatures like the fish. Both in phylogeny and ontogeny, however, the next phase of development is an elongation and folding of the tube to give a U-shaped bend, consisting of the ventricle and the bulbus cordis. Meanwhile, the atrium increases in size, creeps up behind the loop, reaches its definitive position near the cranial end of the pericardial cavity and forms two distinct dilatations, the future *right* and *left atria*. The atrio-ventricular canal now looks ventrally and caudally into the ventricle. The sharp bend from ventricle into bulbus cordis is removed by absorption of the intervening ridge and incorporation of the bulbus cordis into the ventricle. The cranial end of the bulbus cordis and the truncus arteriosus now lie in front of the atrium, pointing towards the head. Gradually this ventriculo-bulbar loop swings over to the left and assumes the adult shape and position.

Inside the mammalian heart the subdivisions of atrium, atrio-ventricular canal and ventricle proceed more or less simultaneously. The *atrio-ventricular (A-V) valve* which prevents blood from returning to the atrium appears first as a series of swellings or *cushions*, projecting into the lumen (Fig. 7.20); these become hollowed on their ventricular aspects to form the usual valvular flaps or cusps which meet in the centre of the lumen whenever the ventricle contracts and fills them with blood.

From the amphibian onwards a streaming or division of the blood flow through the heart is necessary and to achieve this, certain morphological changes are necessary in the atrium, atrio-ventricular canal, ventricle and truncus arteriosus.

The *inter-atrial septum* begins as a crescent-shaped ridge in the roof of the single atrium (Fig. 7.21); when fully developed it corresponds to the *septum primum* of later evolutionary forms. In some amphibia, it grows to form only a partial septum for the atrium while in others it reaches the atrioventricular orifice and directs into the ventricle two functionally distinct streams of blood (one from the lungs, 'oxygenated', and the other 'deoxygenated', from the rest of the body). In later forms, e.g. birds and mammals, subdivision of the atrioventricular canal also occurs by fusion of the anterior and posterior A-V (endocardial) cushions across the lumen to produce right and left A-V canals with cusps developing as described above (Fig. 7.20). In birds the interatrial septum (septum primum) grows towards these fused A-V cushions and fuses with them. This complete interatrial septation is apparently achieved with some stretching of the septum and is accompanied by perforation of its substance re-creating communication between the atria. The foramen (or foramina) heals up after hatching and again completes the septum. In *mammals*, however, the procedure is more elaborate; during fetal life, a *septum secundum* grows from the roof of the atrium on the right of the perforated septum primum but only far enough to cover the gap in the first septum. The two septa do not fuse until after birth but until then, blood passes freely from the right atrium into the

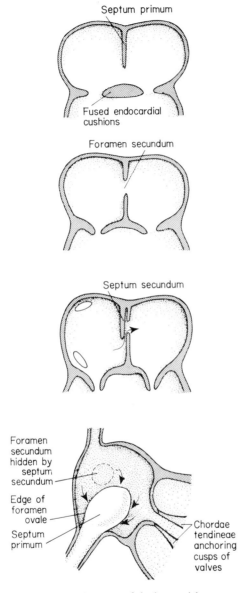

Fig. 7.21 Development of the interatrial septum.

left—and only in that direction—because the first septum yields to the pressure in the right atrium and allows the space between the septa to open up. When, as happens after birth, the pressure in the left atrium comes to equal that in the right, the first septum is pushed against the more rigid second septum and the communication between the two cavities is closed. For further discussion see Fetal Circulation (p.92).

The *inter-ventricular septum* differs in construction and development (Fig. 7.22). Within the early, tiny ventricle, a faint muscular ridge appears at the apex;

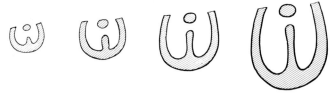

Fig. 7.22 Development of the muscular part of the interventricular septum.

as the right and left ventricular cavities become 'gouged out' of the wall of the heart by the blood streams entering via the A-V canals, this ridge 'stays put' and forms a substantial septum. Because it does not actually grow, this septum cannot separate the two ventricles completely without 'outside' help. In amphibia and reptiles, where complete separation of the blood streams in the heart is not critical, a small opening remains between its free upper border and the fused A-V cushions. In mammals, however, the septum is completed by fibrous tissue derived from the A-V cushions and from the bulbar ridges of the truncus arteriosus (see below).

Blood vessels

The subdivision of the *truncus arteriosus* to provide the *aorta* and the *pulmonary trunk* (and even a third vessel in the crocodile) is achieved by the edge-to-edge fusion of two long *bulbar ridges* which project into the lumen from opposite sides of the vessel (Fig. 7.23). Because these two ridges are spirally arranged along the inside of the truncus, the two new vessels also spiral around one another. Where it arises from the heart, the truncus arteriosus straddles the

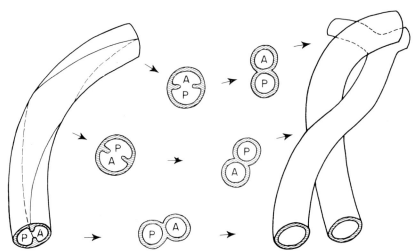

Fig. 7.23 Division of the truncus arteriosus into aorta (A) and pulmonary artery (P) by the spiralling bulbar ridges.

interventricular septum so that it receives blood from both ventricles; and when the bulbar ridges grow back into the ventricular region and fuse with the upper edge of the septum (Fig. 7.24), the blood from the two ventricles is then

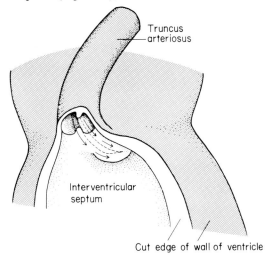

Fig. 7.24 Closure of the opening in the interventricular septum by extensions of growth from the bulbar ridges in the truncus arteriosus. After Hamilton, W.J. and Mossman, H.W., *Human Embryology*, 4th edition. Wm. Heffer & Sons Ltd., Cambridge.

channelled into separate vessels—into the aorta from the left ventricle and pulmonary trunk from the right.

Initially the cranial end of the truncus arteriosus divides to send an *aortic arch* into each half of the first visceral arch. Dorsally, these aortic arches meet to form a single midline *dorsal aorta*. As the remaining visceral arches develop, more aortic arches are provided until there may be six on each side, forming a dorsal aorta on each side before joining the midline aorta (Fig. 7.10). Since the third and succeeding visceral arches become the gill arches in fish, their vessels play a part in gaseous exchange but later evolutionary forms develop major and numerous adaptations in these aortic arches to suit their own needs. An 'average' pattern is shown in Fig. 7.25. A major variation in birds, as compared with most other animals is the persistence of the *right* fourth aortic arch as the *adult aortic arch*.

However, following up the story of the division of the truncus arteriosus into an aorta and pulmonary trunk, we have yet to explain the diversion of blood from the pulmonary trunk into the two pulmonary arteries and the blood in the aorta into the fourth aortic arch. As it leaves the heart the pulmonary trunk lies on the right of the aorta; it then curls round in front of the aorta to lie more or less behind it (Fig. 7.23); at this level the two sixth aortic arches arise from the dorsal part of the truncus arteriosus behind the developing bulbar ridges and order to divert the blood in the pulmonary trunk into the two pulmonary arteries, the two fused bulbar ridges become deflected against and fused with

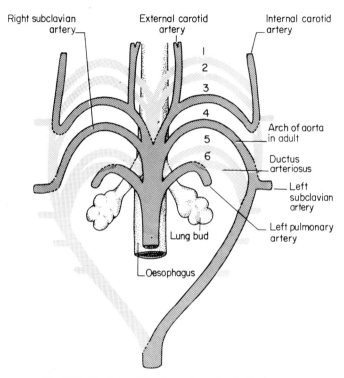

Fig. 7.25 The fate of the six aortic arches in the human.

the dorsal wall of the truncus immediately beyond the sixth aortic arch openings (Fig. 7.26).

From the single dorsal aorta arise numerous branches, (a) some to the neural tube, vertebral column and body wall (the *somatic* group), (b) some to the intermediate mesoderm, pronephros, mesonephros, metanephros and internal genitalia (the *lateral splanchnic*), (c) others to the wall of the gut (the *splanchnic* or *visceral* group) (Fig. 7.27). Later, the segmentally arranged vessels to the gut are reduced in number (Fig. 7.28)—one vessel (the *coeliac artery*) and its branches for the lower end of the foregut and its derivatives, one vessel (the *superior mesenteric artery*) for the midgut but its *vitelline* branches continue alongside the vitello-intestinal duct to reach the yolk sac, and one artery (*inferior mesenteric*) for the hindgut. Since the allantois is an offshoot of the hindgut the *allantoic arteries* also arise from the aorta in this region. In the bird, the branches of these allantoic vessels ramify in the mesoderm of the chorio-allantois but in the mammal, they either follow the same course to reach the placenta or, if the allantois is too small, travel on their own in the umbilical cord to the placenta—in either case bearing the name *umbilical arteries*. The embryo has, in fact, *three* distinct *circulations*, viz. to and from the body, to the *yolk sac* and back and the same for the chorioallantois or *placenta*.

Blood returning from the body is collected by the *anterior* and *posterior*

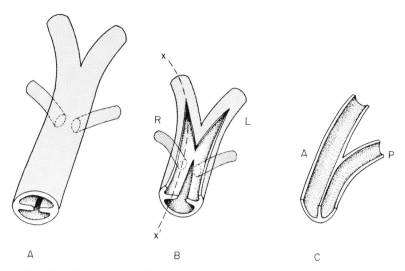

Fig. 7.26 Diagrammatic illustrations to show how the sixth aortic arches become continuous with the pulmonary artery (trunk). (C) represents a section of (B) along the line X—X¹.

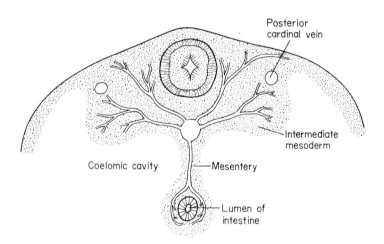

Fig. 7.27 The distribution of the branches of the aorta.

cardinal veins, one of each on either side lying in the mesoderm lateral to the somites (Fig. 7.28). The anterior brings blood back from the head region while the posterior drains the remainder of the body, the two vessels meeting to form the *common cardinal vein* which, lying in the mesoderm of the body wall, swings round the *pleuro-peritoneal channel* to reach the transverse septum and enter the sinus venosus. The common cardinal vein also plays a part in closing off the *pericardial cavity*; when the heart is pushed caudally by the developing visceral arches, the terminal part of the vein moves with it. Instead of the vein

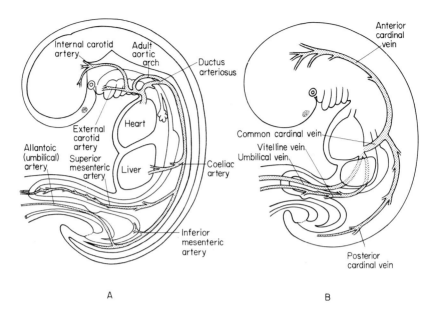

Fig. 7.28 (A), the principal branches of the aorta in the embryo; (B), the principal veins in the early embryo.

maintaining its curved course round the coelomic channel, it is pulled taut and dragged into the channel carrying a shelf of mesoderm with it until it closes the lumen of the channel (Fig. 7.29, 7.30).

In the bird and reptile, the two *vitelline veins*, formed in the mesoderm over the large yolk sac, pass through the umbilicus with the vitello-intestinal duct to meet the veins from the midgut, then plunge into the transverse septum to join the corresponding sinus venosus near the entry of the common cardinal vein. When the yolk is exhausted, the vitelline veins degenerate except for their proximal segments which are retained as the *portal vein* (Fig. 7.31). Even in mammals, where the yolk sac is empty, the vitelline veins develop and behave in the same way.

From the chorioallantois of the chick and reptile and from the allantoic placenta of the mammal the *allantoic* or *umbilical* veins are also large, although the right umbilical vein normally degenerates (Figs. 7.28, 7.31).

Each sinus venosus receives a common cardinal vein and, via the transverse septum, a vitelline and an allantoic (umbilical) vein. However, the developing liver (page 96) breaks up these vitelline and allantoic (umbilical) veins into innumerable sinusoids which thereafter drain into the corresponding sinus venosus by means of a common *hepato-cardiac channel* on each side. Further development (Fig. 7.31) sees (a) the disappearance of the right umbilical vein and (b) the degeneration of the left hepato-cardiac channel. The sinusoids of the liver find it very difficult to transmit all the blood they receive, particularly the blood from the placenta, and therefore, to prevent congestion, one of the sinusoids in the liver opens up to form a large vessel, the *ductus venosus* which

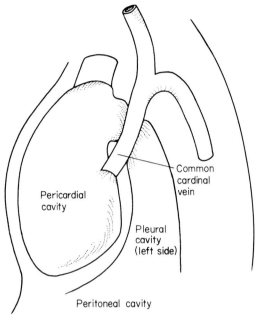

Fig. 7.29 Separation of the pericardial cavity from the future pleural cavity by the common cardinal vein (seen from the left side).

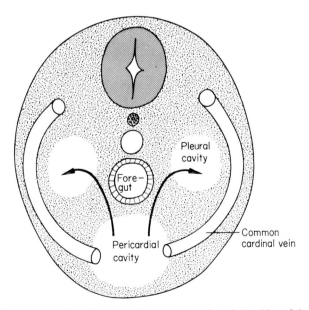

Fig. 7.30 Transverse section through embryo to show the relationships of the pericardial cavity and the future pleural cavity to the common cardinal veins.

short-circuits much of the blood from the left umbilical vein to the right sinus
venosus.

The right sinus venosus is absorbed into the right atrium, and the left
becomes the coronary sinus.

The *pulmonary veins* are entirely new entities in development and cannot be
linked phylogenetically or ontogenetically with any earlier vessels in this region.

The fetal circulation (Fig. 7.32)

From the foregoing, it is obvious that there are special features associated
with the heart and circulation during fetal life in reptiles, birds and mammals,
which use a chorioallantois or a placenta for oxygenation of their blood before
their lungs are functional.

Much of the blood returning from the chorioallantois or placenta via the

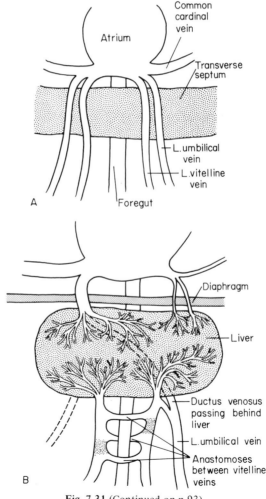

Fig. 7.31 (Continued on p.93)

umbilical vein(s) avoids the sinusoids of the liver by passing through the *ductus venosus* and reaches the heart directly via the inferior vena cava.

Because the blood need not pass through the lungs—the resistance is too high in the pulmonary circulation—much of the blood is diverted through the perforations (bird) or the valvular opening, *foramen ovale*, (mammal) in the inter-atrial septum from the right atrium into the left and thence into the left ventricle. The blood which enters the right atrium from the superior vena cava does succeed in reaching the right ventricle and pulmonary artery but it is diverted into the systemic circulation (aorta) through the ductus arteriosus, the persistent dorsal part of the left sixth aortic arch (mammal) and the dorsal parts of both sixth aortic arches in the bird. The umbilical arteries which arise from the large arteries of the pelvis and leave the body through the umbilicus are also peculiar to the fetus.

In the mammal, when the umbilical cord is ligatured or torn after birth, the umbilical arteries and vein and the ductus venosus collapse and shrivel to fibrous cords; corresponding changes in the circulation occur after hatching when the chorioallantois of the bird and reptile dries up and is discarded. With the onset of breathing, the pulmonary arteries dilate and blood from the right atrium and ventricle reaches the lungs, returning via the pulmonary veins to the left atrium. The ductus arteriosus also constricts and shrivels to a fibrous

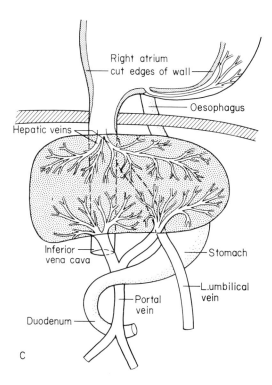

Fig. 7.31 The development of the veins associated with the liver.

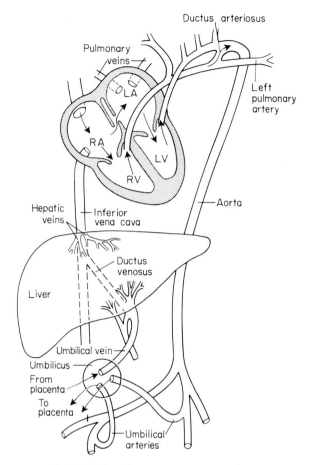

Fig. 7.32 Blood flow in the fetal circulation.

strand. Pressures in the right and left atria are now equal and the two parts of the interatrial septa are apposed and later fused.

THE DIGESTIVE SYSTEM

The development of the *mouth* and *pharynx* was dealt with on pages 75–79, and the development of the oesophagus from the first part of the foregut needs no comment.

Foregut

The remainder of the foregut, which provides the *stomach* and the beginning of the *small intestine*, lies dorsal to the *transverse septum* (Fig. 7.33A), a rather shapeless mass of mesoderm whose development is the key to all that happens in the upper abdomen. In spite of its name it is, initially, only an incomplete septum between thorax and abdomen, because the pleuroperitoneal channels of the coelomic cavity are still present alongside the foregut and its *dorsal* attachment (*mesogastrium*). As the foregut elongates, the *caudal part* of the *transverse septum* becomes drawn out to form the *ventral mesogastrium*, anchoring the stomach to the abdominal wall above the level of the umbilicus. The *cranial part* of the septum is left as a thin transverse sheet—the *diaphragm*

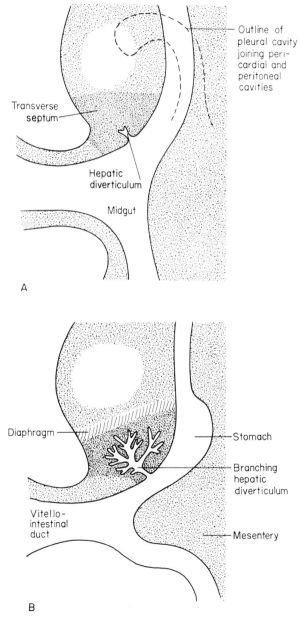

Fig. 7.33 (Continued on p.96.)

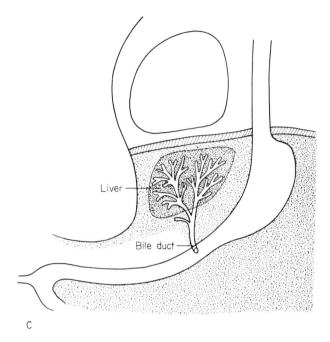

C

Fig. 7.33 Development of the liver and fate of the transverse septum.

(Fig. 7.33B) which, in the mammal, extends dorsally to close off the communication between pleural and peritoneal cavities.

Anchored dorsally and ventrally, the stomach assumes its adult shape by
dilating—mainly dorsally as in the human—but there are more bizarre distortions
—as in the ruminant—all produced by differential dilation or constriction.
The rest of the foregut remains narrow.

Liver (Figs. 7.33C, 7.34)

In the early embyro at the junction of foregut and midgut, a tiny *hepatic
diverticulum* grows into the ventral mesogastrium. As it branches and proliferates vigorously to form the liver, it bulges the mesogastrium both to the right
and left. Close to the stomach, the mesogastrium remains as the thin *gastro-
hepatic (lesser) omentum* while, between the liver and abdominal wall down
as far as the umbilicus, it remains as the *falciform ligament*, the free lower
edge of which carries the umbilical vein.

The liver retains its connection with the gut in the form of the common bile
duct but, in a histological section, the liver shows remarkable little evidence of
its exocrine function. In each lobule, cords of liver cells radiate from the central
vein with wide sinusoids between the cords carrying blood to this vein.
Between the lobules lie the portal canals, each containing a branch of the
portal vein, a branch of the hepatic artery and a small bile duct. These bile
ducts collect the exocrine secretion from the glandular elements, i.e. the cords

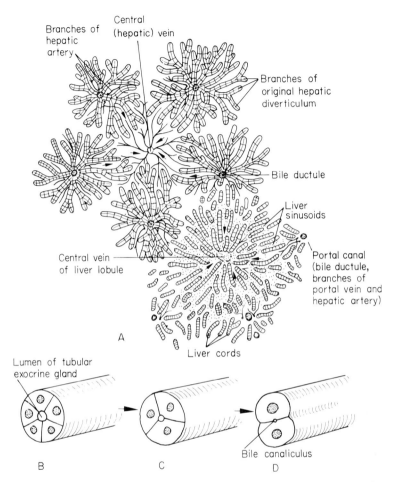

Fig. 7.34 Development of the liver lobule. (A) transformation of 'exocrine' lobules each centred on a bile ductule (in upper part of diagram) to the adult lobule with 'endocrine' functions arranged round a central (hepatic) vein. (B)–(D) development of a liver cord from a terminal branch of original hepatic diverticulum.

of liver cells in the adjacent parts of the surrounding lobules: and the liver cords are simply tubular secretory elements in which the number of cells in each cross-section is reduced to two and the lumen is a tiny (bile) canaliculus draining to the bile duct. The distortion of what was at first an exocrine gland results from the rich and abundant blood supply in the umbilical (allantoic) and vitelline veins being broken up into sinusoids on their way through the liver and from the inevitable trafficking of food and secretions between the liver cells and the blood—an increasingly important function of the organ.

Pancreas (Fig. 7.35)

The *ventral pancreas* in the human grows out in the angle between gut and

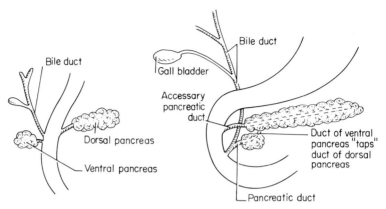

Fig. 7.35 Development of the pancreas.

hepatic diverticulum: at a higher level but from the opposite side of the gut, the *dorsal pancreas* grows into the dorsal mesogastrium. Later the entrance of the bile duct and the ventral pancreas (with its duct) migrate round to the dorsal wall. Here, the two pancreases fuse and the duct of the ventral pancreas 'taps' the duct of the dorsal to carry off its secretion. Thus, both bile and pancreatic secretions enter the intestine together. Minor variations in the development of the pancreas occur in other animals, e.g. in the chick, where the organ begins as two ventral rudiments instead of one, along with one dorsal rudiment. The *pancreatic islets* (*of Langerhans*), which secrete insulin, develop from the pancreatic ducts as tiny buds of cells; they soon lose their connection with the ducts, and become isolated in a sea of acinar tissue, each surrounded by a very thin capsule and permeated by many capillaries.

The upper abdomen is thus divided by a mid-line septum consisting of the ventral mesogastrium (enclosing the liver), the stomach and the dorsal mesogastrium. But the liver usually swings over to the right side and the stomach rotates on its long axis until its original left surface faces forwards. Thus, the part of the peritoneal cavity originally on the right is cut off from the remainder of the cavity except for a narrow connection under the lower edge of the lesser omentum.

Midgut (Fig. 7.36)

The midgut is, at first, the short segment of the gut lying opposite the umbilicus and its ventral wall is incomplete because of the vitello-intestinal duct. Yet from this short piece of gut most of the small intestine and about half of the large intestine are derived.

After closure of the vitello-intestinal duct, the midgut elongates to form a 'loop' of intestine projecting towards the widely open umbilicus; its *mesentery*, meantime, expands to accommodate the growing intestine. The loop has a *cranial limb*, a *caudal limb* and an *apex* marking the attachment of the vitello-intestinal duct, which usually disappears, although occasionally the proximal

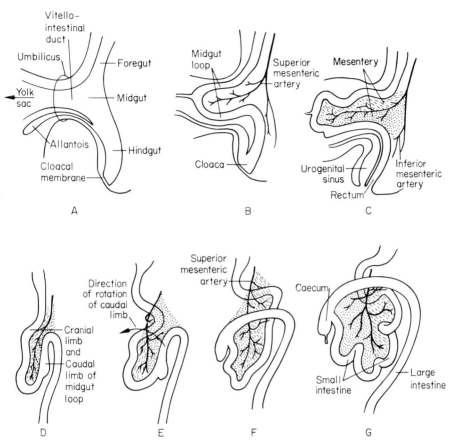

Fig. 7.36 Development and rotation of the midgut. (A)–(C) lateral views; (D)–(G) frontal views.

part of its lumen persists as a small *Meckel's diverticulum.*

On the caudal limb, not far from the apex, the *caecum* begins to grow out. In herbivorous animals, e.g. the rabbit, it forms a long deep sac with a blind end, but in primates it is very short with a shrivelled atavistic distal part, the *vermiform appendix.* The cranial limb and the caudal limb as far as the caecum become the *small intestine* while the rest of the caudal limb provides the *ascending* and *transverse* parts of the colon.

However, with a rapidly growing liver and an abdominal cavity which is slow to increase its size, there is little room for the vigorously developing intestine. The long midgut mesentery fortunately allows it to herniate through the umbilicus into the umbilical cord, where it remains for an appreciable time and continues to grow. When the abdomen can accommodate all its viscera, the midgut loop returns, quite suddenly, in a well ordered and consistent fashion which is fundamentally similar in all animals. Around the superior mesenteric artery as its axis, the whole midgut loop rotates in an anti-clockwise direction when viewed from the front; the upper small intestine passes back through the umbilicus, turning underneath the right side of the superior mesenteric artery to reach the left side of the abdomen pulling the whole cranial limb after it so that the superior mesenteric artery finally lies entirely on the right side of the small intestine. Meanwhile the caudal limb swings over the superior mesenteric artery to lie on its right side; the caecum lies just below the liver with the remainder of the limb draped across the upper abdomen below the stomach and continuous with the hind gut on the left side of the abdomen. Later, caecum and appendix migrate to a position near the pelvis. The end result described here is that found in the human but there are many variations throughout the animal kingdom depending chiefly on the amount of anti-clockwise rotation occurring in the loop as it returns to the abdomen.

Hind gut (Fig. 7.37)

The hind gut is caudal to the level of the umbilicus and, from its ventral surface, the *allantoic diverticulum* grows out towards the umbilicus and thence out into the extra-embryonic coelom. In the angle between allantois and gut, the (splanchnopleuric) mesoderm drives a wedge (the *uro-rectal septum*) between them until it reaches the cloacal membrane. The caudal part of the original hind gut is thereby separated into a *primitive urogenital sinus* ventrally and the rectum dorsally; likewise the two parts of the cloacal membrane, separated by mesoderm from the uro-rectal septum, become the *urogenital opening* and the *anal opening*. In amphibia, birds and reptiles, complete subdivision of the hind gut and cloacal membrane does not occur and the terminal part of the gut is a common excretory chamber (*the cloaca*) for both the urogenital and alimentary systems.

THE RESPIRATORY SYSTEM (Fig. 7.38)

Although by no means typical of all animals and in spite of being evolutionarily advanced, the development of the respiratory system of mammals is described here to provide a pattern from which to gain an understanding of the development and relationship of the other forms.

In the floor of the pharynx in mammals, a longitudinal groove appears; on either side of it a prominent longitudinal ridge projects into the lumen and closes over the groove along its whole length except at the cranial end. Thus, a tubular *lung diverticulum* lying in front of the foregut is created for the future development of larynx, trachea, bronchi and lungs. The *hypobranchial eminence* (Fig. 7.16), lying behind the tongue and in front of this lung diverticulum, becomes the epiglottis, while, around the opening of the diverticulum, the *larynx* develops from the cartilages and muscle of the adjacent visceral arches.

As the diverticulum elongates it branches into two lung buds each of which divides repeatedly to form a bronchial tree. At this stage the lung, embedded in vascular mesoderm, is beginning to project into the adjacent coelomic channel

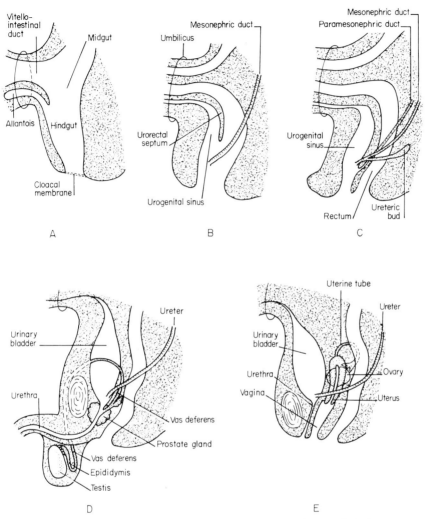

Fig. 7.37 Development of hindgut, urogenital sinus and urogenital ducts. (A)–(C) early stages of development; (D) male genitalia; (E) female genitalia.

which is to become the pleural cavity. Continued growth and division of the lung bud give *bronchi, bronchioles* and finally tiny *alveolar sacs*, all of which are patent from the start and contain fluid; the epithelium of the alveolar sacs has squamous as well as secretory cells from the time they develop and the number of air sacs continues to increase after birth. At birth the fluid is mostly absorbed into the adjacent capillaries to allow air replacement but there is no inflation or distension of the lung at this time.

The evolutionary origin of the mammalian respiratory system can be traced back to fishes while lungs bearing some resemblance to those of the mammal occur in lung fishes. They even arise from the ventral or ventrolateral surface

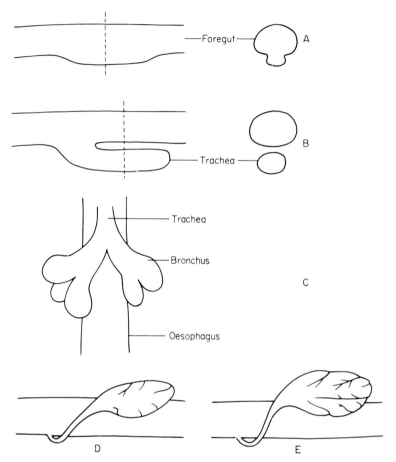

Fig. 7.38 (A)–(C) development of trachea and lung buds. (D), (E) possible evolutionary steps in development of lung.

of the pharynx; septation of these bladder-like structures, as if to give a larger internal surface for the absorption of oxygen, confirms their respiratory function. The *air* or *swim bladder* of other fish is thought to more closely associated with buoyancy although a minor respiratory function cannot be entirely discredited in some cases. The air or swim bladder, believed by some authorities to be the evolutionary precursor of the lung and regarded by others as an evolutionary adaptation of a primitive lung, grows out from the dorsal aspect of the pharynx.

In amphibia and in some lizards, the lung (there may be only one) shows simple septation or subdivision of its cavity while in others, e.g. turtles and crocodiles, the septa projecting into the lung sac are numerous and complicated giving the impression of and an approximation to alveolar sac formation (Fig. 7.38).

Birds probably have the most elaborate respiratory system; large air sacs develop from the main bronchi and, maintaining these connections, invade the neck, abdomen and bones; recurrent bronchi connect these air sacs with the lung tissue proper which has no terminal spaces like those of the mammal. Instead, all the small bronchi in the lungs and all the tiny air capillaries arising from these bronchi fuse with one another at their tips to form loops thus providing for a circulation of air through them. With movement of the wings during flight, the air sacs permeating the body behave as reservoirs of air from which air is pumped through the elaborately designed lungs.

THE UROGENITAL SYSTEM

The two components, urinary and genital, closely linked in the adult, are even more closely associated in development.

URINARY SYSTEM

Lateral to the somites, a similar, but unsegmented, mass of mesoderm stretches along the length of the embryo; this is the *intermediate mesoderm* (Fig. 7.39) which projects into the coelom as a long urogenital ridge on each side of the mesentery and gives rise to nearly all the urogenital system.

Pronephros (Fig. 7.39)

The pronephros develops in the intermediate mesoderm of the cervical region, first as an aggregation of mesenchymal cells in each 'segment' to form a solid cord which later canalizes to give a *pronephric tubule*; one end of the tubule opens into the coelom (it may even begin as an invagination from the coelom into the intermediate mesoderm) while the other end turns caudally to join with the tubule in the next segment. Repetition of this behaviour results in a series of transversely arranged tubules, linked by a longitudinal *pronephric duct*. A cluster of capillaries (the *glomerulus*) may project either into the coelom close to the opening of the tubule or into the tubule itself. This description of the development of the pronephros applies to most vertebrates but a variation in the mode of development of the pronephric duct is described for the amphibian; here it is thought to arise as a thickening of the surface cells of the intermediate mesoderm, subsequently elongating to form the lumen of this duct.

Mesonephros (Fig. 7.39)

Except in the most primitive creatures, the pronephros disappears but the intermediate mesoderm of the whole thoracic and lumbar region succeeds in forming a more efficient (and longer lasting) excretory system, the *mesonephros*. A *mesonephric tubule* develops on either side for each body segment, elongates and becomes intensely convoluted. In some fish and amphibia, the tubule may

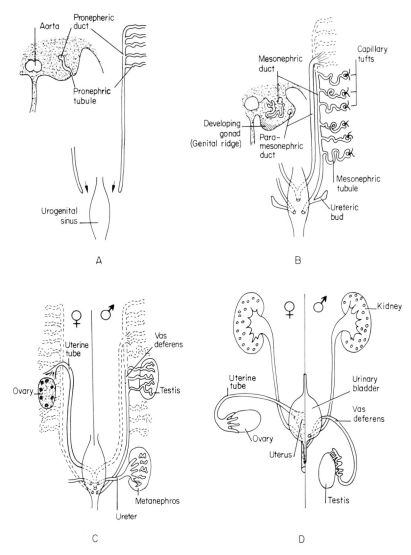

Fig. 7.39 Development of the urogenital system in the male and female embryo.

open temporarily into the coelom but in other vertebrates no connection develops at any stage. In all forms one end, the 'promixal' end, of the mesonephric tubule becomes invaginated by a developing glomerulus to form a *'corpuscle'* from which a transudate from the blood passes into the mesonephric tubule. Although the pronephros itself degenerates, its duct, now called the *mesonephric duct*, continues to grow caudally through the territory of the mesonephros where the other, 'distal', end of each mesonephric tubule joins it and uses it as a collecting duct for the 'urine' filtered from the blood. This duct continues to grow caudally until it reaches the cloaca; when the cloaca is completely subdivided by the urorectal septum the mesonephric ducts open into the urogenital sinus (Figs. 7.37, 7.39). Some lower forms, e.g. the frog, retain the caudal part of the mesonephros as functional tissue into adult life; it is incorporated into similar nephrogenic material developing in the cloacal region and corresponding to the metanephros of birds and mammals (Fig. 7.40).

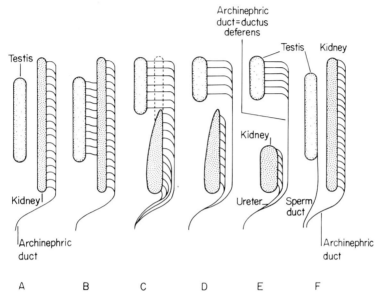

Fig. 7.40 Phylogenetic variations in the urinary and genital ducts and in the development of the kidney. From Romer, A.S., *The Vertebrate Body*, 2nd edition. W.B. Saunders Company, Philadelphia.

In birds and mammals the mesonephros degenerates during fetal life although some of its components are salvaged for the male genital system.

Metanephros

A third type of kidney develops in reptiles, birds and mammals and persists as their permanent kidney. The mesonephric duct, as it enters the urogenital sinus, produces a little offshoot, the *ureteric bud* (Figs. 7.37, 7.39), which grows cranially and dorsally as the ureter and plunges into the intermediate mesoderm in the pelvic region: there it branches repeatedly to provide the collecting duct system, i.e. pelvis of ureter, major and minor calyces and collecting tubules, of the *metanephros* or true kidney. When it enters the metanephros, the ureteric bud and every one of its branches becomes 'capped' with a dense mass of mesoderm. Each cap forms a solid cord of cells and then a tubule, one end of which becomes invaginated by a glomerulus of capillaries (as in the mesonephros) while the other joins the end of the collecting tubule. The rapid elongation of the metanephric tubule provides the proximal and distal convoluted tubules and the loop of Henle. Those parts of the tubular system derived from metanephrogenic mesoderm comprise the *nephron*. The kidney later migrates cranially to a position opposite the lumbar vertebrae.

The caudal ends of the mesonephric duct and the ureteric bud, are absorbed into the urogenital sinus until they open into it separately (Fig. 7.39).

THE GENITAL OR REPRODUCTIVE SYSTEM

The primordial germ cells, which give rise to the ova and spermatozoa, can be identified first, not in the gonads, but among the endodermal cells of the yolk sac; from there they travel to a region on the medial surface of the mesonephros forming a distinct swelling (the *genital ridge*) (Figs. 7.39, 7.41) which becomes eventually either the testis or the ovary.

Testis (Fig. 7.41)

In the developing *testis*, *sex cords*, consisting of primordial cells and mesenchymal cells appear and proliferate deep in its substance. At a later stage— perhaps not until the time of sexual maturity—these cords become canalized as the *seminiferous tubules*, containing in their walls the *Sertoli cells* derived from mesenchyme and the *spermatogonia* from the primordial germ cells. At one side of the testis, the sex cords run together to form an anastomosing meshwork, the *rete testis*, also canalized later. Throughout the reproductive period of life, the spermatogonia acting as stem cells divide by mitosis to give *primary spermatocytes*: these divide by meiosis to give haploid *secondary spermatocytes*; next come the *spermatids* and finally the *spermatozoa* which are anchored to or embedded in the Sertoli cells for a time before 'casting off' into the lumen. Further details of the process of spermatogenesis can be found in the following chapter. Between the tubules lie the *interstitial cells*, arising from the mesoderm and responsible for the endocrine secretion of the testis.

Ovary (Fig. 7.41)

In the developing *ovary*, clusters of cells (follicles) rather than cords, each comprising a central germ cell surrounded by mesenchymal cells develop in the peripheral part (cortex) of the organ. Oogenesis (development of the ovum) and the accompanying phenomena leading to ovulation (discharge of the ovum from the Graafian follicle) are described in the next chapter. The mature ovum, still surrounded by some of the follicular cells, is released into the peritoneal (coelomic) cavity. In birds, where the ova contain large amounts of yolk, the developing and mature follicles project from the surface of the ovary on vascular pedicles.

Genital ducts and external genitalia (Figs. 7.37, 7.39, 7.42)

Some duct system must develop to 'retrieve' the ovum from the peritoneal

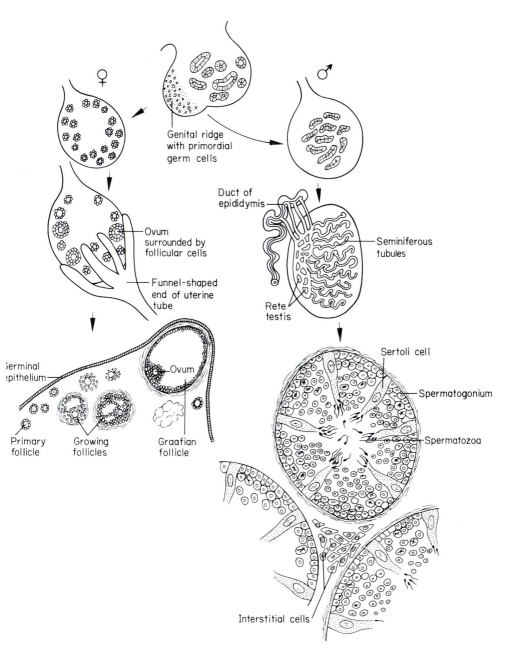

Fig. 7.41 Development of ovary and testis; oogenesis and spermatogenesis.

cavity. This is provided for, at least potentially, in the embryo of either sex. Thus, opposite the gonad, an invagination from the coelomic cavity burrows into the lateral side of the mesonephros and runs caudally alongside the

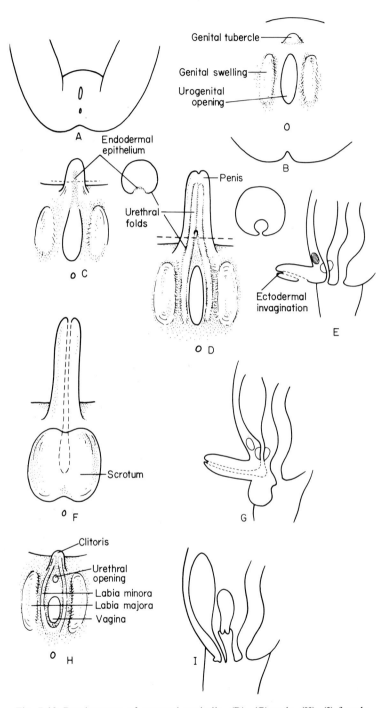

Fig. 7.42 Development of external genitalia. (D)–(G) male. (H), (I) female.

mesonephric duct as the *paramesonephric duct*. In the pelvis, the two paramesonephric ducts join one another behind the urogenital sinus and enter it as a single tube.

In the male embryo, the paramesonephric ducts disappear but, in the female, they persist; the cranial end, retaining its opening into the peritoneum, flares out like a funnel with a fimbriated edge to 'catch' the ova as they are released from the ovary. In primates, the remainder of the unfused portions become the uterine tubes, the fused part the uterus and vagina; but in many cases the unfused portions each form a uterus as well as a short uterine tube, with the fused part becoming the vagina. The intermediate condition of a Y-shaped uterus is also common.

Phylogenetically the male genital system possesses no means of its own for transporting the spermatozoa to the surface of the body. Primitively, these were shed into the coelom and extruded through abdominal pores but in higher forms (birds and mammals), the rete testis which drains the seminiferous tubules opens into the mesonephric tubules and duct which are adapted as *epididymis* and *vas deferens* to carry the products to the cloaca or urogenital sinus.

In those species where the urogenital and digestive systems have a common cloaca, i.e. fish, amphibia and birds, there may be minor modifications at or near the cloacal opening for copulation. In some reptiles and in mammals generally a distinct pattern exists in the form and development of the external genitalia. With the degeneration of the allantois, the cranial end of the primitive urogenital sinus is sealed off. That part of the sinus cranial to the entry of the ureters dilates to become the urinary bladder while the caudal part, the *definitive urogenital sinus*, develops as if it were two distinct portions—a cranial and a caudal. In each sex, the *cranial portion* remains a narrow tube; in the male it becomes the part of the urethra which receives the mesonephric ducts (later the ejaculatory ducts) and from which the numerous glandular elements of the prostate develop; in the female where it is demarcated inferiorly by the entry of the fused paramesonephric ducts, it forms the whole urethra. The caudal portion of the definitive sinus elongates dorsiventrally and becomes flattened from side to side; in the male it helps form the rest of the urethra (see below); in the female, it forms the vestibule with the urethra and the fused paramesonephric ducts (vagina) opening into it.

The early development of the *external genitalia* is also the same in both sexes. Around the opening of the flattened and elongated part of the definitive urogenital sinus three mesodermal masses arise—one on each side, the *genital swellings*, and one in front, the *genital tubercle*. In the *male*, the endodermal sinus epithelium migrates some distance over the ventral surface of the genital tubercle; along the sides of this epithelial sheet, two *urethral folds* arise and, as in neural tube development, meet to form a tube. The folds are continued back alongside the urogenital opening between it and the genital swellings where they also close over in the mid-line. Thus the lumen of the sinus is led into the newly formed tube in the genital tubercle (penis); at the tip of the penis an ectodermal pit grows in to link up with and complete the urethra. In the female, development in this region does not progress so far: the genital tubercle (*clitoris*) remains small but has endoderm on its ventral surface; the 'urethral' folds do not fuse but form the *labia minora* leaving the sinus to open on the surface as the *vestibule*: the genital swellings form the *labia majora*.

The migration (descent) of the gonads (Figs. 7.43, 7.44)

Neither the testis nor the ovary remains in its original position on the dorsal aspect of the abdomen. A long narrow condensation of mesoderm attached to the gonad, runs (caudal) down the abdominal wall, round the brim of the pelvis and through the ventral abdominal wall to anchor itself in the genital swelling. This, the *gubernaculum testis* or *ovarii*, (depending on the sex) guides or pulls the gonad down behind the peritoneum from its original position. In front of the gubernaculum testis as it pierces the abdominal wall, a peritoneal pocket (*processus vaginalis*) also protrudes into the genital swelling (scrotum). The testis and epididymis, trailing its vas deferens and blood vessels, slip down behind the pocket and invaginate it. The neck of the pocket is later obliterated. The migration of the testis does have its variations. In some animals no definitive scrotum develops and the testis is located in a kind of pocket of the peritoneal (coelomic) cavity. Even when a scrotum is present, the testis is not in it all the time. The testis can slip down into the scrotum and back into the abdomen according to the season of the year: for instance, in some rodents during the breeding season it moves into the scrotum where the temperature is lower—apparently a prerequisite for spermatogenesis.

Only in the primates does the gubernaculum ovarii have much influence on the position of the ovary, but because of the attachment of the gubernaculum to the paramesonephric duct (uterine tube), it pulls the ovary only as far as the pelvic cavity. In other species the ovary and uterine tube are moved only a little from their original position and remain in the abdomen attached to the dorsal wall.

The pattern of urogenital development described here is, in the phylogenetic sense, an advanced one but its basic features were established as early as in the invertebrates e.g. the segmental repetition of nephric units (see page 205). The primitive pattern and its successive adaptations can often be traced, adding interest to the subject without seriously invoking the principle that ontogeny recapitulates phylogeny. The opening of pronephric tubules from the coelomic cavity reflects the early excretory mechanism of invertebrates, supplemented later by capillary tufts in the coelom and followed by the invagination of these capillaries (glomeruli) into the tubules; nor has the use of every segmental excretory unit been abandoned in the vertebrates e.g. in the hagfish embryo, where the kidney (holonephros) extends the whole length of the intermediate mesoderm.

In the genital system, too, there are interesting correlations between ontogeny and phylogeny. Primitively, spermatozoa and ova were all discharged into the coelom and escaped via abdominal pores; an interesting intermediate phylogenetic

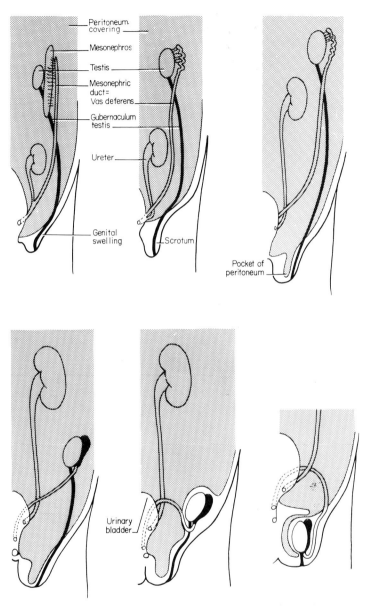

Peritoneum covering

Mesonephros

Testis

Mesonephric duct = Vas deferens

Gubernaculum testis

Ureter

Genital swelling

Scrotum

Pocket of peritoneum

Urinary bladder

Fig. 7.43 The descent of the testis.

stage is seen in Fig. 7.40 where the urinary excretory duct is shared by kidney and testis and it could be said that the evolution of the male genital system reflects a continual struggle to achieve a separate excretory pathway to the exterior—only partially successful to date.

In the urogenital system and elsewhere in the body there are many other

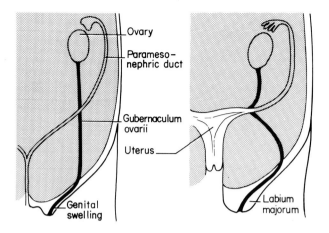

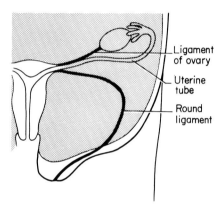

Fig. 7.44 The descent of the ovary in the human.

similar intriguing and debatable features—all helping towards a better under-standing of growth and development.

Hormonal differences in the development of the genitalia

For the gonad, the decision to become testis or ovary lies with the sex chromosomes but the mechanism of their action is still obscure. Development of the genital ducts and the external genitalia according to sex is directed by the gonad; again the mechanism is not clear but in both cases much experimen-tal investigation has been applied to the problem. In general, the female condition of gonads and genital ducts may be regarded as the neutral or basic state and maleness is superimposed on it. The administration of male hormones (androgen) to female fish and amphibian embryos changes them to functional males; likewise female hormones administered to male embryos produces functional females but this is not the case with mammalian embryos. Experiments involving castration, the administration of androgens after castration and the administration of an 'anti-androgen' to mammalian embryos have given interesting results from which the following deductions may be made: the developing testis secretes two 'hormones'; one, which can be 'neutralized' by anti-androgen and is therefore an androgen or of a similar nature, stimulates the mesonephric duct to develop into epididymis and vas deferens while the other 'hormone', unrelated to androgen inhibits the development of the paramesonephric duct. Other experiments involving removal of one fetal testis or the implantation of fetal testis to one side of a female fetus demonstrate that the hormones secreted by the fetal testis have a 'local' effect only, i.e. only on the genital ducts of that side. It is very doubtful if the ovary has a comparable endocrine effect because removal of the ovary causes no change from the normal female pattern of development.

A long dormant period in the development of the genital system usually exists almost until sexual maturity when the testis and ovary reach full activity; this is achieved by the pituitary gonadotrophic hormones stimulating both the production of spermatozoa and maturation of ova as well as the gonadal endocrine secretions i.e. androgen and oestrogen. The latter, in turn, are responsible for the secondary sex characteristics of the animal.

FURTHER READING

Arey L.B. (1966) *Developmental Anatomy* 7th ed. Philadelphia & London: W.B. Saunders Company.

Assali N.S. (1968) *Biology of Gestation* Vol. 2. New York & London: Academic Press.

Balinsky B.I. (1970) *An Introduction to Embryology* 3rd ed. Philadelphia & London: W.B. Saunders Company.

Barnes A.C. (1968) *Intra-uterine Development.* Philadelphia: Lea & Febiger.

DeHann R.L. & Ursprung H. (1965) *Organogenesis.* New York & London: Holt, Rinehart & Winston.

Hamilton W.J. & Mossman H.W. (1972) *Human Embryology* 4th ed. Cambridge: W. Heffer & Sons Ltd.

Kraus B.S., Hironori K. & Lathan R. (1966) *Atlas of Developmental Anatomy of the Face.* New York & London: Harper & Row.

Marrable A.W. (1971) *The Embryonic Pig.* London: Pitman Medical.

Nelsen O.E. (1953) *Comparative Embryology of the Vertebrates.* London: Constable.

Patten B.M. (1968) *Human Embryology* 3rd ed. New York & London: McGraw-Hill Book Company.

Romanoff A.L. (1960) *The Avian Embryo.* New York: The Macmillan Company.

8 GAMETOGENESIS AND FERTILIZATION

The primordial germ cells, once established in the ovary, are known as *oogonia*, the stem cells of the ova. In mammals, oogonia, by a series of mitotic divisions, proliferate to produce primary oocytes in enormous numbers; it has been estimated for example that there are as many as seven million oocytes in the human embryo by the fifth month. In many species, all these mitotic divisions are complete by or just after birth and the oocytes proceed to undergo meiosis. The first meiotic division however is arrested in each oocyte at a time which varies according to the animal and each oocyte then enters a dormant phase characterized in amphibia and in birds, by an increase in size and by the accumulation of yolk in its cytoplasm. Also during this phase there is proliferation and thickening of the follicular cells around the oocyte but, with accumulation of yolk in the ovum, the envelope of follicular cells becomes stretched and thin. Around the mammalian ovum which has very little yolk and therefore does not increase in size as markedly as other ova, the follicular cells proliferate to produce many layers of cells; later, fluid collects amongst the cells and a cyst develops with the ovum pushed to one side still surrounded by follicle cells (Fig. 7.41).

After an animal reaches sexual maturity, it experiences breeding seasons or oestrous cycles which vary considerably in frequency according to the species from once a year to regularly every few days. Each cycle or season is accompanied, in those animals which have a significant amount of yolk in the ova, by an acceleration in the rate of deposition of yolk in the ova which are to be released in that cycle; the number of ova may be quite small or there may be hundreds or thousands as in fish and amphibia. In mammals where the yolk content of the ovum is insignificant, the final stages of ovum maturation are marked by a rapid increase in the volume of fluid in the follicular cavity. In both cases the follicles with the ova protrude on the surface of the ovary, rupture and allow the ova to escape. Another important feature associated with the final stage in the maturation of the ovum is the completion of the first meiotic division, release of the first polar body, followed by the second meiotic division as far as the metaphase stage. Often but not invariably the completion of the second meiotic division and release of the second polar body await fertilization.

The exact number of oocytes maturing in each cycle seems to depend on the level of the Follicle Stimulating Hormone (FSH) in the blood stream and on the sensitivity of the ovary to that hormone. The release of the ovum (ovulation) is brought about by the action of Luteinizing Hormone (LH).

Many features of oogenesis are still unexplained. For instance, are all the oocytes in the ovary derived in early fetal life from the initial settlement of primordial germ cells or oogonia, or is there fresh production of oogonia and oocytes in adult life from ovarian tissue such as the surface (germinal) epithelium? Nor is there any explanation of why the number of oocytes should be so high in the early fetal ovary of the mammal relative to the number actually ovulated, or of why the number of oocytes in the ovary should fall so rapidly in later fetal life and in the young immature animal.

In the developing testis (Fig. 7.41), the primordial germ cells, now regarded as *spermatogonia*, and the accompanying mesenchymal-type, *sustentacular (Sertoli) cells* are arranged in the form of cords. These cords eventually become seminiferous tubules although the lumen of the tubule is not obvious until the animal is sexually mature. Between the tubules, *interstitial cells* develop from the mesenchyme to supply the endocrine secretions responsible for the secondary sex characteristics. Little or no activity of the spermatogonia occurs until, with the approach of sexual maturity, they begin mitotic proliferation and initiate spermatogenesis, a process which may continue into old age. While some spermatogonia remain as stem cells, others become *primary spermatocytes* and these by meiotic division produce *secondary spermatocytes*. By this time, the number of germ cells far exceeds that of the sustentacular cells. The spermatocytes differentiate into *spermatids*, smaller cells containing small dark-staining nuclei and partially embedded in the cytoplasm of the Sertoli cells. Spermatids in their transition to *spermatozoa* (Fig. 8.1) undergo major structural changes; the nucleus condenses even more; an *acrosomal vesicle* or *acrosome*, containing hydrolytic enzymes like hyaluronidase, materializes from coalescence of the small vesicles of the Golgi apparatus and positions itself over the future 'front' end of the nucleus; the centrioles approach the nucleus on the side opposite to the acrosome and from one of them grows the flagellum of the spermatozoon; the mitochondria of the cell aggregate around the proximal part of the flagellum, and the rest of the cytoplasm is sloughed off before or just after the spermatozoon is released into the lumen of the tubule.

When the ova and spermatozoa are discharged simultaneously into the surrounding water (e.g. in amphibia) fertilization occurs outside the body; in birds and mammals, however, the spermatozoon meets the ovum in the upper end of the uterine tube and fertilization occurs there. Most ova are enclosed in enveloping vitelline membranes which the spermatozoon must penetrate by enzymic action. To achieve this (Fig. 8.2), there is, first, fusion of the wall of the acrosomal vesicle in the head of the spermatozoon with the overlying plasma membrane; breakdown of the area of fusion releases the hyaluronidase from the vesicle to create a path for the spermatozoon through the vitelline membrane. When the spermatozoon reaches the ovum, the posterior wall of the acrosomic vesicle (now acting as the plasma membrane over the leading surface of the spermatozoon) fuses with the plasma membrane of the ovum; the fused

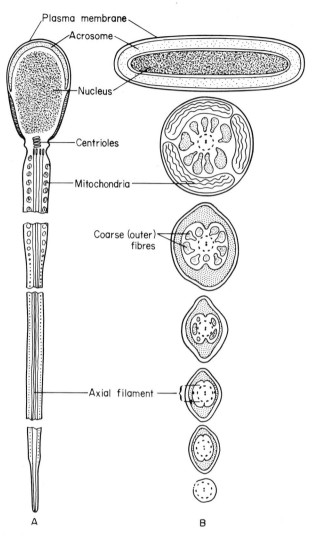

Fig. 8.1 The structure of the spermatozoon; (A) in longitudinal section; (B) (at a higher magnification) in a series of transverse sections. From Austin, C.R., *Fertilization*: Foundations of Developmental Biology Series. Prentice Hall Inc., Englewood Cliffs, N.J.

membranes then break down and the cytoplasms of the two gametes, ovum and spermatozoon, become confluent. In fact, the haploid spermatozoon nucleus accompanied by its mitochondria and centrioles all enter the ovum; the tail usually remains outside and degenerates. A *fertilization membrane* develops from granules in the outermost layer (cortex) of the egg cytoplasm which become detached from the surface of the ovum and applied to the vitelline membrane; at the same time a new plasma membrane is developed to enclose the ovum, or, as it is now known, *the zygote.*

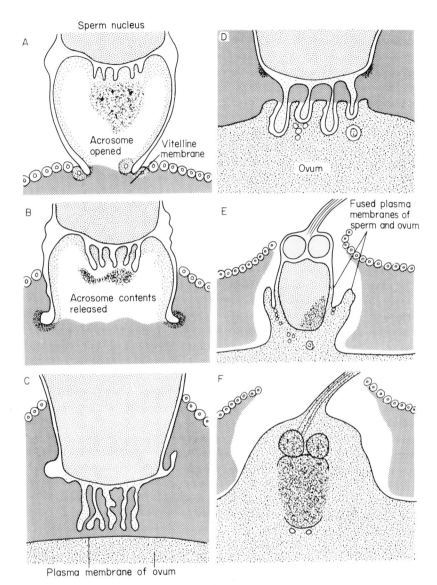

Fig. 8.2 Fertilization; cytoplasmic fusion of spermatozoon and ovum. From Ebert, J.D. and Sussex, I.M., *Interacting Systems in Development*, 2nd edition. Holt, Reinhard and Winston Inc., New York, London. (After A.L. Colwin and L.H. Colwin.)

Meanwhile, with the entry of the sperm contents, the arrested meiotic division of the nucleus in the ovum starts up again, completes the nuclear division and the second polar body is extruded, leaving the other nucleus or *female pronucleus* inside. The sperm nucleus swells to become the *male pronucleus*. Each of the centrioles entering the ovum from the spermatozoon produces a daughter centriole and then, accompanied by the newly formed centriole, they separate from one another. At the same time, the chromosomes in the two pronuclei coil and shorten and the surrounding nuclear membranes disappear; thereafter all the chromosomes aggregate and form a metaphase plate between the two centrospheres. The process continues as if it were a simple mitotic division resulting in two diploid nuclei. Cleavage of the rest of the zygote follows if the content of yolk will allow it.

Recent research work has revealed that these features are only a part of the activation of the egg; for instance, there is also an awakening of the metabolic and synthetic mechanisms. During oogenesis, stores of food accumulate in the cytoplasm, sometimes in enormous quantities (e.g. in birds). To what extent this was the result of synthesis within the ovum, or synthesis in and transfer from surrounding follicular cells, or simply transfer from maternal tissues through the follicular cells is not yet clear but RNA synthesis is a conspicuous feature in the developing ovum and all the cellular organelles necessary for protein synthesis are present in the ovum. By the time the ovum is mature, both RNA and protein synthesis are in abeyance but after fertilization protein synthesis is activated (or reactivated). In the amphibian, but not in the mouse, this immediate post-fertilization synthesis of protein is surprising because RNA synthesis in the amphibian ovum does not resume until the early stages of gastrulation; the explanation lies with the excess production of ribosomal and messenger RNA occurring during oogenesis and persisting in a 'masked' or inactive form (as a protein complex) until fertilization. The exact mechanism of unmasking of the RNA after fertilization has not been established.

This is only a sketch of the intricate processes occurring before, during and immediately after fertilization and although there is an increasing literature on these topics, much has still to be learned before a complete understanding is possible.

FURTHER READING

Austin C.R. (1961) *The Mammalian Egg.* Oxford: Blackwell Scientific Publications.

Austin C.R. (1965) *Fertilization* Foundations of Developmental Biology Series. Englewood Cliffs, N.J.: Prentice-Hall, Inc.

Austin C.R. & Short R.V. (1972) *Reproduction in Mammals* Vol. 1 *Germ Cells and Fertilisation.* Cambridge: University Press.

Betz C.B. & Monroy A. (1967) *Fertilisation* Vol. 1. New York & London: Academic Press.

Berrill N.J. (1971) *Developmental Biology.* New York & London: McGraw-Hill Book Company.

Biggars J.D. & Schuetz A.W. (1972) *Oogenesis.* London: Butterworths.

Cole H.H. & Cupps P.T. (1969) *Reproduction in Domestic Animals* 2nd ed. New York & London: Academic Press.

Parkes A.S. (1960) *Marshall's Physiology of Reproduction* 3rd ed. Vol. 1 Pts 1 and 2. London: Longmans.

Romanoff A.L. (1960) *The Avian Embryo.* New York: The Macmillan Company.
Spratt N.T. Jr. (1971) *Developmental Biology.* Belmont, California: Wadsworth
    Publishing Company, Inc.

9 THE OESTROUS CYCLE AND
ENDOCRINOLOGY OF REPRODUCTION

In mammals, the release of ova from the ovary is preceded, accompanied and followed by a complex series of morphological and endocrinological changes in the body. The pattern of these changes varies with the species and is known as the *oestrous cycle* or (in the human and some higher primates) the *menstrual cycle.*

THE OESTROUS CYCLE

For several days before the Graafian follicle ruptures, its growth has been under the control of the *Follicle Stimulating Hormone* secreted by the pars distalis (anterior) of the pituitary gland. But this hormone also stimulates the secretion of a steroid hormone, *oestrogen*, by the ovary, particularly from the theca cells surrounding the follicle: oestrogen, in turn, causes proliferation of the epithelium and underlying connective tissue of the endometrium (inner layers of the uterus). When the oestrogen reaches a sufficiently high level, it suppresses the secretion of the FSH and the pars distalis begins secreting the *Luteinizing Hormone.* Its function is to cause rupture of the Graafian follicle and release of the ovum and thereafter to stimulate the secretion of a second steroid hormone, *progesterone*, from the follicular cells. The accumulation of the yellow lipid material in these cells and their proliferation results in the forma-tion of the *corpus luteum.* Progesterone increases the vascularity of the endo-metrium and induces its glands to secrete. If no fertilization of the ovum ensues, the corpus luteum degenerates, the production of LH, oestrogen and progesterone ceases and that oestrous cycle ends.

The oestrous cycle of the rat lasts 4—5 days, during which the Graafian follicle grows under the influence of FSH: the amount of LH is sufficient only to rupture the follicle but insufficient to form a corpus luteum. The endo-metrium merely regresses to its earlier condition and a new cycle with fresh FSH secretion begins. *Fertilization* of the ovum is accompanied by the secretion of enough LH to form a corpus luteum and hence the secretion of progesterone to prepare the endometrium for implantation of the blastocyst. Mating without fertilization may, however, initiate the same series of changes to give a false or psuedo-pregnancy which may last a few days and include enlargement of the mammary glands. All these features regress when the corpus luteum degenerates. It appears, therefore, that mating is, in itself, sufficient stimulus, acting upon the hypothalamus and pituitary gland for the secretion of more luteinizing hormone, enough to cause corpus luteum formation as well as ovulation. In the rabbit, oestrogen secretion seems to be more or less continuous, bringing a series of Graafian follicles up to maturity, but allowing several to remain in that state for a period of time which overlaps the period when the next batch of follicles are mature. Thus several mature follicles are always present in the ovaries. Without mating, no follicles rupture because no LH is released by the pituitary. Even a sterile mating, however, stimulates enough LH secretion to cause ovulation and to form a corpus luteum with an ensuing psuedo-pregnancy. In the bitch, the oestrogenic phase is followed naturally by rupture of the follicle and, particularly after an infertile mating, there is a pseudo-pregnancy which may persist for several weeks before all the features regress.

MENSTRUAL CYCLE (Fig. 9.1)

In the human cycle there is always a pre-ovulatory and a post-ovulatory phase controlled by the FSH/oestrogen and LH/progesterone respectively; the latter is characterized by increased vascularity and oedema of the endometrium and increased secretion by the uterine glands. Apart from the LH/progesterone phase following inevitably upon the FSH/oestrogen phase, there is a new feature in the cycle of the human and of some other primates, i.e. sloughing and discharge of most of the endometrium (*menstruation*) whenever the

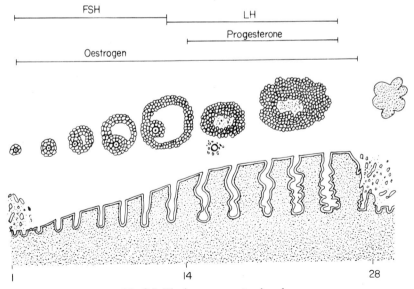

Fig. 9.1 The human menstrual cycle.

progesterone secretion is cut off on degeneration of the corpus luteum; the natural occurrence of this discharge in the human approximately every 28 days is the reason for calling it the *menstrual cycle* (L. menstruus, monthly). The second half of each menstrual cycle, controlled by the secretions of the corpus luteum and accompanied by some increase in size and engorgement of the mammary glands is therefore comparable to the pseudo-pregnancy of the bitch.

The control of the oestrous cycle

A more striking variation in oestrous cycles is their frequency—ranging from those which follow one another without interlude throughout the year, e.g. in the rat, to those which occur once or twice a year e.g. in the bitch, and always at approximately the same time of the year. Such control over the initiation of the cycle as well as the effect of mating on the production of LH by the pituitary, e.g. in the rabbit, suggests that external stimuli or environmental factors such as day-length may influence the cycle and, in fact, it has been shown that the pituitary gland is under the control of the hypothalamus lying immediately above it. Specific releasing factors for FSH and LH are produced in the hypothalamus and transported by a portal system of veins to the pars distalis and, although the nervous pathways whereby visual and other sensory impulses from the environment reach the hypothalamus have not been clearly defined, there is good reason to accept the hypothesis.

PREGNANCY

The endocrinology of pregnancy is an elaboration of the conditions prevailing in the oestrous cycles and, chiefly because it has been the most extensively investigated, the human pregnancy will be used here to illustrate the main features; do not extrapolate freely to other species however. The continuing increase of oestrogen throughout the menstrual cycle and the increase in progesterone during the second phase continue into pregnancy. Numerous closely allied oestrogenic substances have been identified but the main feature is their enormously raised level in the maternal tissues and in the urine. Initially secreted by the corpus luteum which degenerates after a few months, oestrogen production is soon supplemented by and finally taken over by the trophoblast of the placenta. The placenta does however require the co-operation of the fetal and probably maternal tissues as well for the synthesis of the oestrogens. Trophoblast is also responsible for maintaining and increasing the secretion of progesterone but still another hormone makes its appearance and figures prominently in the endocrinological picture associated with the human pregnancy. *Human chorionic gonadotrophin* (HCG), a protein hormone, was first identified in the urine of pregnant women and its presence there has been the basis of pregnancy tests for many years. It, too, is synthesized by the trophoblast and its level in the maternal tissues rises sharply from the early days or weeks of pregnancy. The role of HCG is initially to prolong the life and secretory activity of the corpus luteum by supplementing the production of gonadotrophins by the pituitary gland. In this way the corpus luteum of menstruation persists as the corpus luteum of pregnancy and continues to produce oestrogen and progesterone. At or about the third month of pregnancy, the secretion of HCG has reached its peak; thereafter it drops but the hormone is present in appreciable concentration until the end of pregnancy. Its origin from trophoblast is emphasized by the fact that it continues to be excreted if trophoblast should be left in the uterus after the fetus and rest of the placenta are delivered.

Coinciding with the fall in HCG production about the third month, the corpus luteum begins to degenerate and the levels of oestrogen and progesterone now secreted by the trophoblast maintain the uterus in a condition which will nourish and safeguard the conceptus; but they have other functions, e.g. altering the metabolism and biochemistry of the mother so that she may better nourish the fetus.

The *mammary gland* during pregnancy is prepared for lactation by the action of oestrogen and progesterone; these hormones increase the branching of the duct system and stimulate the development and growth of the secretory alveolar portions but do not initiate the secretion of milk. This occurs under the action of *prolactin* secreted by the pars distalis of the pituitary: without preliminary exposure of the mammary glands to the steroids however, the prolactin has no effect. Continued suckling by the child can maintain the secretion of prolactin for many months and the same stimulus is also responsible for the secretion, from the pars nervosa of the pituitary, of *oxytocin* a hormone which causes contraction of the walls and emptying of the ducts of the mammary gland.

It is surprising how little influence these maternal hormones have on the growth and development of the fetus. Only in certain organs is there any evidence of effect in the newborn, e.g. a watery secretion from the mammary gland (witch's milk), temporary enlargement of the uterus and prostate gland and changes in the vaginal epithelium, all of which could arise from stimulation by steroid hormones. Early in fetal life and continuing even until a few months after birth, the interstitial cells of the testis are well differentiated with signs of secretory activity—probably an HCG effect.

FURTHER READING

Assali N.S. (1968) *Biology of Gestation* Vol. 1. New York & London: Academic Press.

Austin C.R. & Short R.V. (1972) *Reproduction in Mammals* Vols. 4, 5 & 6. Cambridge: University Press.

Barrington E.J W. (1963) *An Introduction to General and Comparative Endocrinology.* Oxford: Clarendon Press.

Cole H.H. & Cupps P.T. (1969) *Reproduction in Domestic Animals* 2nd ed. New York & London: Academic Press.

Josimovich J.B., Reynolds M. & Cobo E. (1974) *Lactogenic Hormones, Fetal Nutrition and Lactation.* New York & London: John Wiley & Sons.

Klopper A. & Diczfalusy E. (1969) *Foetus and Placenta.* Oxford: Blackwell Scientific Publications.

Nalbandov A.V. (1964) *Reproductive Physiology* 2nd ed. San Francisco & London: W.H. Freeman & Company.

Parkes A.S. (1960) *Marshall's Physiology of Reproduction* 3rd ed. Vols. 1, 2 & 3. London: Longmans.

Perry J.S. (1971) *The Ovarian Cycle in Mammals.* Edinburgh: Oliver & Boyd.

10 GROWTH AND AGEING

Everyone knows what they mean by growth but few would care to define it precisely. Increase in height or length, increase in weight, increase in cell size, in cell numbers or even in intercellular substance could all result in growth and, indeed, depending on the circumstances, have been used as a means of measuring it. Although growth is important in all animals the human has received far more attention in this respect and the discussion here will deal mainly with people, treating growth under three broad headings; first the growth of the body as a whole, then growth of the organs and tissues and finally some reference to the growth process as it affects the cell.

THE GROWTH OF THE BODY

Human growth is an individual matter, some people growing more quickly than others, some becoming taller, some remain thin while others become fat and so on but there is nevertheless a basic pattern for everyone. Even a casual glance at a series of human embryos and fetuses (Fig. 10.1), arranged according to age, reveals that there has been rapid growth before birth but closer study shows that, in spite of the marked increase in size over these nine months, the *rate* of increase has *decreased* towards the end of gestation. Comparison of the size of the newborn child with a one-year old and then with a two-year old makes it clear that there is a slowing up of growth over these two years. In other words, the successive percentage increments of growth per day, per week or per month diminish steadily throughout fetal and postnatal life. The progression is not entirely smooth because just before or just after birth there may be an even slower rate of growth compensated later by a period of increased growth. Until 8–10 years of age the child nevertheless continues to grow but at an ever decreasing rate, becoming so slow that it may appear to have stopped although this is unlikely. A brief spurt (or increased rate) of growth occasionally occurs around 6–7 years of age but about 11–12 years, often later, there is an obvious and definite recrudescence of growth, the *adolescent growth spurt* (Fig. 10.2), lasting for about 1½–3 years when the child may gain several inches in height; thereafter the height gradually approaches its adult level and shows no increase

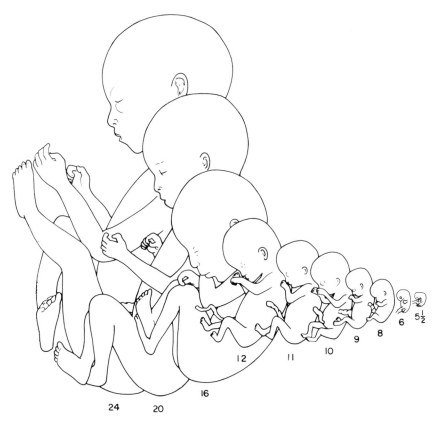

Fig. 10.1 Growth of the human fetus between $5\frac{1}{2}$ and 24 weeks gestation; $\times\frac{1}{3}$. After Patten, B.M., *Human Embryology*, 3rd edition. McGraw-Hill Book Company, New York, London.

beyond 25 years of age. In the mature individual the ensuing changes are not usually associated with growth because there is no increase in height, but they are consistent and predictable i.e. increased weight and changes in the profile of the adult, a gradual thickening of the soft tissues, a broadening of the frame and, by 'middle age', an increase in the amount of fat deposited in some regions particularly in the abdomen. If these changes are unacceptable as part of the picture produced by growth, then naturally they come under the category of ageing which could be considered as the sequel to growth. It is important to remember that the adolescent growth spurt is not a conspicuous feature (if it occurs at all) in lower animals; and, in the fish, growth appears to continue throughout its whole life with only minor seasonal fluctuations, mainly associated with feeding and spawning.

Variations in the human growth pattern are as important as the pattern itself. The onset of the adolescent growth spurt, for instance, is particularly variable; girls generally begin their growth spurt sooner than boys (Fig. 10.3), a feature best seen in the case of boy-girl twins (Fig. 10.4); the girl's early

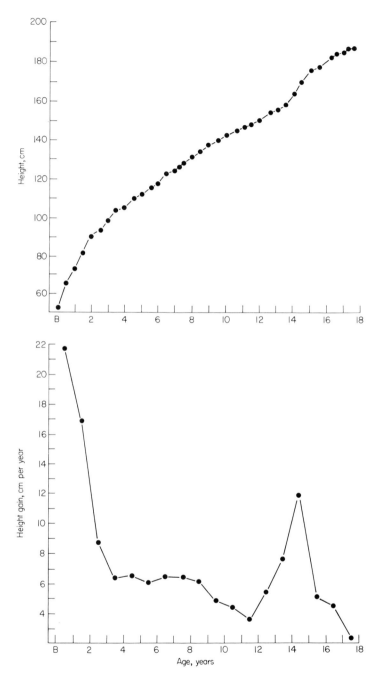

Fig. 10.2 Growth in height in a boy from birth to 18 years (1759–1777). Above; distance curve, height attained at each age. Below; velocity curve (rate of growth), increments in height from year to year. From Tanner, J.M. *Growth at Adolescence*, 2nd edition. Blackwell Scientific Publications, Oxford.

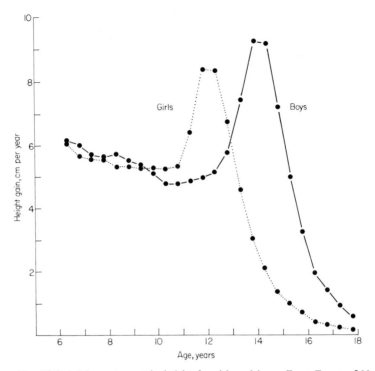

Fig. 10.3 Adolescent spurt in height for girls and boys. From Tanner, J.M., *Growth in Adolescence*, 2nd edition. Blackwell Scientific Publications, Oxford.

increase in height contrasts with her brother's delay; later, perhaps when she has almost ceased growing, he catches up and soon outstrips his sister in height and weight. This difference between boys and girls corresponds to the difference in the time of onset of puberty between the sexes. Individual variation in the onset of the growth spurt is also a marked feature; some boys, beginning early, may have completed it before some of their class-mates have shown signs of even beginning it. The increased physical development accompanying the increased height may be reflected in the complications in psychological and intellectual development of school children in the same class or group.

The earlier onset of puberty in both boys and girls to-day, compared with children of 50–100 years ago, is accompanied by an earlier growth spurt and it follows that, on average, the children of the 1970's are taller than their grandparents were at equivalent ages. Not only does the adolescent reach his or her full height earlier but they are also ½–1" taller when they reach it compared with their ancestors of 100 years ago. These features have been discussed and debated for many years in an effort to explain them but the improvement in general nutrition and social conditions is regarded as the most likely cause. The suggestion that the adult height and weight, which are largely determined by the genetic make-up of the individual, have been or are being

Fig. 10.4 Twins showing the different growth rates in the boy and girl. The intervals between the white lines, the same in all four photographs are the length of the boy's head (crown to chin) and are a rough indication of its varying proportion to his stature. From Lockhard, R.D., *Living Anatomy*, 6th edition. Faber and Faber, London.

gradually altered over the generations by changes in the genes themselves to give taller men and women finds less support. An important implication therefore of these observations is that the level of nutrition plays an important role in the rate of growth and perhaps in the amount of growth attained by the individual.

TISSUE AND ORGAN GROWTH

Throughout the growing period, i.e. from fetal life to adulthood, the various parts of the body do not grow at the same rate or at the same time. At birth the proportionate sizes of head, trunk or limbs are quite different from what prevails in the adult and in relation to the adult size of these parts (Fig. 10.5). The skull of the newborn, which reflects the growth of the brain, is large compared with the trunk and limbs and at 5 years of age has reached about 90 per cent of its final size. The face, including the lower jaw, on the other hand grows slowly during childhood, accelerating during puberty to reach its full size. The limbs, also proportionately small at birth grow fairly steadily during childhood but during the adolescent growth spurt, the legs, particularly, grow rapidly and indeed account for most of the increased growth in height at

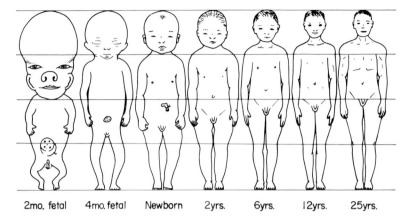

2mo. fetal   4mo. fetal   Newborn   2yrs.   6yrs.   12yrs.   25yrs.

Fig. 10.5 Two fetal and five postnatal stages of development drawn to the same total height to show the changes in the proportions of the various parts of the body. From Patten, B.M., *Human Embryology*, 3rd edition. McGraw-Hill Book Company, New York, London.

that time. Variation is just as marked in other parts of the body; the early growth spurt of the nervous system before birth results later in a relative shortening of the spinal cord compared with the vertebral column which grows steadily throughout childhood and adolescence (Fig. 10.6). The liver which also grows rapidly in the early fetus, still occupies relatively more of the abdomen in the newborn than in the adult. Perhaps the most unusual growth 'curve'

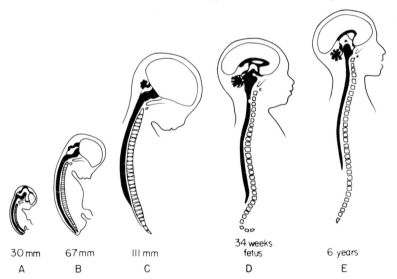

30 mm   67 mm   111 mm   34 weeks   6 years
                                      fetus
  A   B   C   D   E

Fig. 10.6 The changes in the level of the caudal end of the spinal cord at different stages of fetal life. (A), (B) and (C) from Hamilton, W.J. and Mossman, H.W., *Human Embryology*, 4th edition. Wm. Heffer & Sons Ltd., Cambridge, (D) and (E) from Patten, B.M., *Human Embryology*, 3rd edition. McGraw-Hill Book Company, New York, London.

belongs to the genitalia; laid down and partially differentiated in fetal life, they grow no more until the approach of puberty when they not only complete the process of differentiation, e.g. in the seminiferous tubules, but also grow rapidly during puberty. Lymphoid tissues e.g. thymus, tonsils, are in marked contrast; steady growth in childhood until puberty is followed by gradual atrophy thereafter. There are good reasons for understanding these variations in the growth curves of the different organs but explanations of the mechanisms involved are less easy to find. The growth spurts associated with puberty are, however, initiated by the hormones emanating from the pars distalis of the pituitary gland and the production of these is now recognized as coming under the influence of the hypothalamus. The gonadotrophic hormones and the changes occurring at puberty have been described on p.121 but these hormones also act with the somatotrophic (growth), thyrotrophic and adrenocorticotrophic hormones during the growth spurt. These and particularly the somatotrophic hormone are responsible for the skeletal growth at the epiphyseal plates but exactly how the growth hormone has such a specific effect on these tissues at the ends of a bone is not clear.

Variations in the time of onset of the adolescent growth spurt should be mentioned here in relation to bone growth. Each adolescent may be reckoned as having a *chronological age*, i.e. the number of years he or she has lived but, as we have seen, it is not an accurate guide to the rate or stage of growth (*biological age*). Skeletal growth follows a remarkably consistent pattern as far as the relationships in the appearance and fusion of the different ossification centres. These features usually follow one another in the same order but the whole pattern is more closely related to a child's biological age than his chronological age. A child's *skeletal age*, defined as the stage of bone growth and development is therefore a better indication of the growth rate, the likely onset of puberty and the timing of the growth spurt in terms of years. Details of the skeletal age are best seen on radiographs of the adolescent bones and the standard region for ascertaining the skeletal age is the left wrist (Fig. 10.7). The timing of tooth eruption (Fig. 10.8), establishing a *dental age*, is also a better guide than the chronological age as to when puberty and adolescent growth are likely to occur.

GROWTH AT THE CELLULAR LEVEL

This is more difficult to investigate but is none the less important. Mitosis, of course, is essential for growth but control of the growth process (including initiation and cessation) as well as differentiation must be exercised at the molecular level and the mechanisms should be sought there. In large measure, we then become involved in the problems of differentiation, looking for methods whereby the genome may be differentially stimulated or inhibited to provide the characteristics of the particular tissue. In relation to fluctuations of growth such as the adolescent spurt, hormones could be regarded as evocators but it is

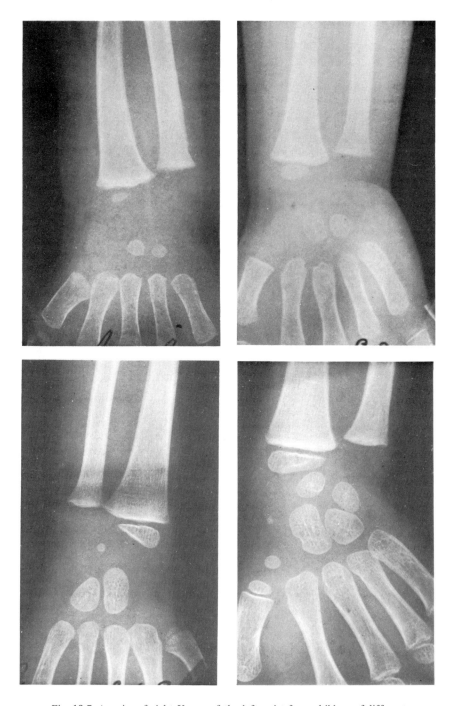

Fig. 10.7 A series of eight X-rays of the left wrist from children of different

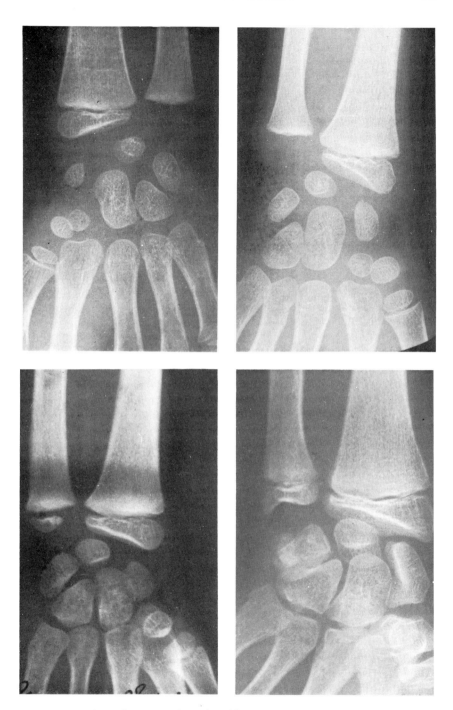

ages to show the progression of ossification in the bones of the wrist. (By permission of Mr. A.M. Stewart.)

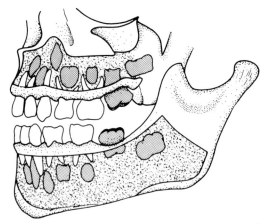

Fig. 10.8 The teeth of a child about six years of age. The outlines of the permanent teeth are finely stippled. From *Gray's Anatomy*, 33rd edition; Edited by Davies, D.V., and Davies, F. Longmans.

unlikely that one hormone is entirely responsible; almost certainly many factors, even several hormones, are involved in establishing an environment in which the main hormone can initiate the next step in the growth process. Control of mitotic division may be achieved in several ways, e.g. via an effect on gene activity, on the processes of protein synthesis or on the general metabolism of the cell. It has been suggested, from experimental work with serial cell cultures, that the cells of some tissues can undergo only a finite number of mitotic divisions and when these are completed, no more are possible; no satisfactory explanation can be given for this feature. There is also evidence that mitotic division is controlled by *chalones*; these substances, produced by the cells themselves, inhibit mitosis when their concentration rises in the cytoplasm and environment of the cells. With reduction, loss or escape of these chalones, which are tissue specific, mitotic division will resume to provide the number of cells necessary to bring the concentration of chalones back to an inhibitory level again. Here, therefore, is a mechanism whereby a tissue or organ may limit its own size. We must not, however, overlook the influence of nutrition on cell growth; obviously, ample food and other nutrients are likely to facilitate all the metabolic processes of a cell and thus accelerate its growth. Severe limitation of these resources will cause a gross disturbance of cell growth but moderate inhibition may have the effect of slowing growth and multiplication of the cells to a degree which is obvious only when measured over a long period. This explains the later onset of puberty and of the adolescent growth spurt and the slower attainment of the adult genetic height when the child's level of nutrition is less than the optimal. It may also be reasonable to suppose that a high level of nutrition can accelerate growth and perhaps cause overshoot of the expected height and weight.

AGEING

Like growth, ageing is an enigma but even more so. The two processes overlap but cessation of growth is as difficult to define as the onset of ageing. In the adult, once he has reached a certain age, it is easy to identify the signs of old age though not all of them need be present, e.g. loss of hair pigment (greying), dryness and wrinkling of the skin, stooping and loss of height, spots of pigmentation in the skin, loss of agility, increased weight due to fat, poor vision and forgetfulness. In spite of the variety of signs and symptoms, little is known with certainty about the process of ageing although many suggestions have been made to explain it and each may have some truth in relation to particular aspects.

The proposal that the cells of a tissue have only a finite number of mitotic divisions implies that those cells may have reached their finite number by the time the tissue or organ is fully grown. This, combined with the view that a cell must either divide or die within a certain period, i.e. it is not immortal, is responsible for the view that in some tissues, e.g. the brain, there is a progressive loss of cells from the time that increase in numbers ceases. In the case of the nervous system this theory seems particularly applicable since mental acuity and memory do deteriorate and there are fewer nerve cells in old age.

Another widely accepted view of ageing sees changes in the intercellular substances as the most important element. Collagen acquires increased cross-linkages in its protein molecules, while elastic tissue loses its elasticity with the passage of years. Thus there is a hardening and loss of resilience in dense connective tissue and cartilage. Changes in the amorphous intercellular material e.g. an increase in the amount of chondroitin sulphate, also seem to be partially responsible for the ageing of tissues. We should also consider the progressive increase in the collagenous intercellular material in tissues such as skin and realize that such increase alone may have an effect on the vascularity, resilience and general nutrition of the tissue.

The tendency to increased deposition of fat in the older person occurs so often that there appears to be a diminution in the ability to metabolize fats, with the result that they are simply stored in different parts of the body if the intake is not restricted. This reminds us that failure of other metabolic processes and the associated failure of synthetic activities may also be part of ageing and be responsible on the one hand for the loss of pigment in hair and on the other for the development of small pigmented areas in the skin of the face and arms. Pigment is also found in the cytoplasm of different cells, e.g. heart muscle, as they age.

Ageing is often manifest in the form of degenerative disease. In joints, arthritis arising from degeneration of the cartilage, and reactionary calcification around the joint surfaces is a common, painful, although innocuous accompaniment of age. When the process affects the arteries, however, it is more serious; degeneration and disappearance of the elastic tissue in the tunica media of the blood vessels (arteriosclerosis), followed in some cases by calcification results in weakness and loss of resilience in the walls. Often accompanying arteriosclerosis is a degeneration of the intimal lining of the blood vessels, particularly in the coronary arteries of the heart, and followed usually by the appearance of fatty deposits or plaques (atherosclerosis) projecting into the lumen. Irregularity and roughening of the intima in this way form the prelude to thrombosis or blood clotting occurring in the coronary arteries.

There is some justification for believing that loss of cells and degeneration of the tissues may result from mutations occurring spontaneously during mitotic division in the cells and sometimes caused by ionizing irradiation. Experimental work on animals has not always supported this argument however. The mutations are believed to result in abnormal intercellular tissues, abnormal secretions or the abnormal by-products of disordered metabolism. Associated with this view is an explanation of the delay in the appearance of these abnormal features of somatic mutation until the later years of life; it is suggested that in the earlier years the immunological responses of the body succeed in removing or destroying the abnormal products or even the cells themselves but, with the decline of the immunological responses (cf. the diminution of lymphoid tissue from adolescence onwards), these abnormal cells and their abnormal functions persist and manifest themselves as defects in the tissues.

More people are nowadays reaching a ripe old age and beyond, but this achievement is probably due to the avoidance of intercurrent illness such as infection or to more successful treatment, e.g. of cancer. Ageing however seems inevitable and proceeds in spite of all attempts to halt or inhibit it and it is not enough to say that its onset and progress are genetically determined. Observations and experiments do suggest one way of slowing or inhibiting the process. Animals which have been fed particularly well and on a highly nutritious diet succumb earlier than their counterparts on a spare diet. When we realize that better nutrition and improved living conditions have been shown to accelerate the growth of animals and man, it is reasonable to suggest that the acceleration and subsequent earlier cessation of growth allow for an earlier onset and more rapid progression of ageing, evidenced by degenerative diseases such as arterio- and athero-sclerosis of the blood vessels. Any attack on ageing may therefore be more successful if it begins with a prolongation of the process of growth achieved by providing an optimum rather than a maximum diet. It may therefore be too late to resort to 'dieting' or 'slimming' in middle age in order to prolong life.

FURTHER READING

Cheek D.B. (1968) *Human Growth*. Philadelphia: Lea & Febiger.
Falkner F. (1966) *Human Development*. Philadelphia & London: W.B. Saunders Company.
Finch C.E. (1969) *Cellular Activities during Ageing in Mammals*. New York: MSS Information Corporation.
Harrison G.A., Weiner J.S., Tanner J.M. & Barnicot N.A. (1964) *Human Biology*. Oxford: The Clarendon Press.

Illingworth R.S. (1970) *The Development of the Infant and Young Child* 4th ed. Edinburgh: E. & S. Livingstone.

Kohn R.R. (1971) *Principles of Mammalian Ageing.* Englewood Cliffs, N.J.: Prentice-Hall, Inc.

Ounsted M. & C. (1973) *On Fetal Growth Rate.* London: William Heinemann Medical Books Ltd.

Rockstein M. (1973) *Development and Aging in the Nervous System.* New York & London: Academic Press.

Rockstein M. & Baker G.T. III (1972) *Molecular Genetic Mechanisms in Development and Aging.* New York & London: Academic Press.

Smith D.W. & Bierman E.L. (1973) *The Biologic Ages of Man.* Philadelphia & London: W.B. Saunders Company.

Spratt N.T. Jr. (1971) *Developmental Biology.* Belmont, California: Wadsworth Publishing Company, Inc.

Strehler B.L. (1962) *Time, Cells and Ageing.* New York & London: Academic Press.

Symposium of the Society for Experimental Biology (1967) *Aspects of the Biology of Ageing.* Cambridge: University Press.

Tanner J.M. (1961) *Education and Physical Growth.* University of London Press.

Tanner J.M. (1962) *Growth at Adolescence* 2nd ed. Oxford: Blackwell Scientific Publications.

Teir H. & Rytomaa T. (1967) *Control of Cellular Growth in Adult Organisms.* London & New York: Academic Press.

Thomson D.A.W. (1942) *Growth and Form.* Cambridge: University Press.

Winick M. (1972) *Nutrition and Development.* New York & London: John Wiley & Sons.

11 CELL DIFFERENTIATION

How can one cell, the fertilized ovum, provide as its progeny an infinitely large number of cells which, apparently without guidance or outside interference, differentiate from one another to become the cells of brain, of muscle, of liver, of bone and so on? This is the process of *differentiation* which, although still not properly understood, is fundamental in developmental biology.

In the amphibian embryo we can see a neural plate appear, then the development of a neural groove and neural folds, and if we go on watching it, the folds close over the groove and the neural tube disappears below the surface to become the brain and spinal cord. In sectioned embryos at different ages, we can identify each step in the development—the thickened neural plate with its tall elongated cells, the stage of the neural folds, the closure of the folds and separation of the neural tube from the original layer of cells, i.e. the ectoderm; later there are the more complicated stages leading to the formation of a brain and spinal cord. Corresponding steps in the development of all other organs and systems can also be easily identified and in each case there are visible changes in the size, shape, staining reactions, arrangement and other features of the cells.

These are gross, relatively late changes in a cell; and they could not have occurred without preceding alterations or modifications in the metabolic and synthesizing mechanisms. If we take, as an example, the differentiation of the muscle cell, it is obvious that the synthesis of muscle proteins must be initiated long before there is any visible sign of differentiation and continued until these proteins form the main element in the cytoplasm to the exclusion of those proteins and secretions which are characteristic of other kinds of differentiated cells. One may assume that each cell, no matter its fate, originally has the potential to synthesize all proteins characteristic of the organism, i.e. that all cells possess all the genes of that animal. In a few exceptional cases however, somatic cells are known to lose genetic material—even whole chromosomes, e.g. in some insects and nematodes. Differentiation therefore usually entails selecting the genes responsible for the synthesis of particular proteins whether they are to remain in the cell as enzymes or as the 'structural', e.g. muscle, proteins or to be discharged as a glandular secretion. There must also be an accompanying suppression or inactivation of the genes which are not required in a cell.

COMPETENCE AND DETERMINATION

The above argument does little more than set out the problem of differentiation and, indeed, there is little we can say with any conviction about how differentiation is initiated and maintained in a cell. However, a great deal of patient observation and brilliant experimental work has been brought to bear on the problem and we must review the information obtained to date.

In normal development, tissues and organs consistently derive from the same layers of the early embryo, i.e. ectoderm, endoderm, or mesoderm or from the same combinations of these. However, in experiments on the amphibian embryo, cells can be removed from one embryonic layer, e.g. ectoderm, and transplanted to another, e.g. endoderm; if done at the blastula stage, the transplanted cells settle down and differentiate in the same fashion as the surrounding endodermal cells; they are then said to be *competent* to change their pathway of differentiation. If the same procedure is attempted at the gastrula stage or after, the transplanted ectodermal cells do not fall into line with their new neighbours but continue to develop as ectodermal derivatives; their competence has become much less and they are becoming *determined* as far as their fate is concerned. Within the ectodermal layer, however, it is possible during the gastrula stage to transfer cells from the presumptive skin area to the presumptive or definitive neural plate area and still find that the cells show sufficient competence to become neural cells. True, the later this transfer is performed during gastrulation, the less likely the presumptive epidermal (skin) cells are to be transformed or differentiated into neural cells and by the end of gastrulation all competence to do so is lost. In other words, the ectodermal cells show decreasing competence combined with increasing degrees of determination in regard to their pathway of differentiation as development proceeds. This is a simple example illustrating the progressive nature of differentiation. Differentiation may be regarded as a pathway which divides repeatedly; at first there is one common path for all cells then three groups go their separate ways, as ectodermal, endodermal and mesodermal cells; and the further they travel the more difficult it is for them to turn back and try another route. Fig. 11.1 takes this analogy even further. All cells differentiate like this, becoming more and more individualist as differentiation proceeds.

MECHANISMS IN CELL DIFFERENTIATION

Other features of differentiation will come to light when topics like induction, regeneration and cell culture are discussed but on the subject of differentiation itself there is still the problem of how it occurs. With every embryonic cell having the potential in its genome to become any type of adult cell, the search for its causes and mechanisms must lie in the environment of the cell. It is easy to visualize different effects on the cell arising out of differences in its location in the embryo. In the chick embryo, for instance, the early segregation of cells

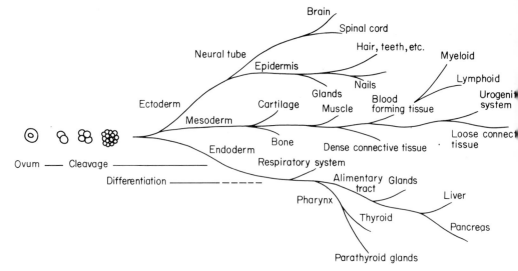

Fig. 11.1 A scheme to illustrate the diverging pathways of cell and tissue differentiation.

into epiblast and hypoblast must lead to differences however minute in the availability of food (yolk) and of oxygen, the hypoblast having the advantage in the former case and the epiblast in the latter. These features are bound to lead to differences in the metabolism of two kinds of cells which are otherwise similar: thus, if different substrates and metabolites are presented to similar cells, then not only will the enzymes involved in dealing with them be different but the products of their activities will also be different and, more important, the genes involved will be different too. Although the differences are slight and few in the case of the embryonic disc giving rise to few changes in the activity of the genome, the segregation of cells into epiblast and hypoblast is only a first step, to be followed by others, in the subsequent differentiation of cells in both epiblast and hypoblast. Another example of position or environment influencing cells at a critical stage occurs in the development of trophoblast from the outer layer of the mammalian morula and the inner cell mass within it giving rise to the embryo proper. Determination in these cells is established very early and, at that stage, probably the only difference between the two groups is the relative ease or difficulty with which they acquire food and oxygen from the uterine secretions.

Ooplasmic segregation

In the amphibian ovum the unequal distribution of yolk, concentrated at the vegetal pole, and cytoplasmic organelles, found mostly in the animal pole, must be reflected in the composition of the cells after the third (equatorial) cleavage. Not only are there four micromeres sitting on top of four macromeres but the amount of yolk material particularly in relation to the number of active cytoplasmic constituents such as mitochondria, ribosomes and lysosomes must be strikingly different and could influence the rate and the type of enzymic reactions in the two types of cells. We could expect then that the nature of the products of intracellular activity would be very different. Intercellular diffusion and interactions between these products soon extends the range of variation in nucleocytoplasmic reactions in the cells—all stemming from the original ooplasmic segregation of material.

Induction

Such a mechanism, i.e. ooplasmic segregation, can only be an effective means of initiating differentiation in the early stages of development in those ova with moderate or large quantities of yolk, but differentiation also occurs in embryos developing from ova with minimal quantities of yolk and continues into later stages of development. Under these circumstances another important mechanism is responsible for much of differentiation, namely *induction*, a process which is defined as the means whereby one group or type of cells in the embryo causes a second group to differentiate along quite a different pathway from that on which they had originally embarked and different from that of the inducing cell. There are numerous examples. In the amphibian, the invaginating presumptive notochordal cells in the dorsal lip of the blastopore come to lie underneath the area of ectoderm known as presumptive neural tissue although, in fact, it is the presence of the notochordal cells which stimulates the overlying ectoderm to differentiate into the neural plate. Experimentally this has been easily demonstrated; presumptive notochordal cells from the dorsal lip of the amphibian blastopore were taken from one gastrula and placed in the blastocoel of another at the same stage, and in such a position that they lie underneath the ectoderm at some distance from the 'natural' blastoporal lip. The result was not only the development of a new, second, notochord in the gastrula but also the differentiation and development of a new neural plate and tube; in fact there is eventually a second complete embryonic axis. The explanation of how it is achieved is still in doubt so many years after it was recognized that the noto-chordal cells were having this effect. Even the question of whether cell contact occurs between the two kinds of cell is unanswered because it is proving difficult in spite of interposing millipore filters and examining the cells by electron microscopy, to show whether contact exists or not.

A more popular avenue of research into this problem has been the investigation of whether induction was dependent on intercellular transfer of a specific chemical substance and, in view of the hypothesis that it is the (micro-) environment of the nuclear material which is responsible for stimulating and/or repressing the relevant genes, there was every likelihood of a successful outcome to the search. Unfortunately a profusion of substances have been 'identified' as inducing agents, e.g. nucleic acids, glycogen, steroids, cephalin, but the first disturbing revelation was that killed notochordal cells could induce neural tissue; later came the demonstration of induction by adult tissue living or dead and finally the fact that induction could occur after the application of inorganic

material, such as methylene blue, to the ectoderm. The idea that there was or should be a specific inducing substance for neural tissue was difficult to defend thereafter. Since then the lac operon theory has been invoked to explain differentiation by induction. According to that theory (Fig. 11.2), a

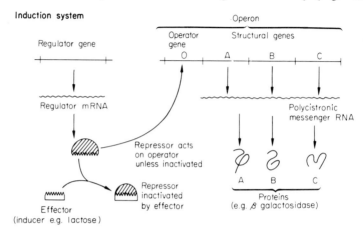

Fig. 11.2 The lac operon system. From Truman, D.E.S. *The Biochemistry of Cytodifferentiation.* Blackwell Scientific Publications, Oxford.

regulator gene produces a substance which inhibits the action of the operator gene; the operator gene when not inhibited in this or in any other way initiates the activity of the structural genes which are responsible for the production of a particular enzyme or other protein. In the bacillus *Escherichia coli*, the (repressor) substance produced by regulator gene activity combines with lactose whenever this sugar enters the cell; but, in doing so, the repressor substance loses its inhibitory effect on the operator gene and the structural genes proceed to produce the enzyme β-galactosidase which metabolizes the lactose. When the sugar is completely metabolized, the repressor substance, now free, reverts to exerting its inhibitory effect on the operator gene and, once more, synthesis of β-galactosidase ceases. Although this is a reversible reaction in unicellular organisms, and in fact has only been satisfactorily demonstrated in these, it has been postulated as the mechanism whereby an inducing substance activates irreversibly a previously repressed part of the genome in embryonic cells of a multicellular organism. The non-specific feature of the inducing agent is explained by suggesting that a variety of substances, organic or inorganic, could have a neutralizing effect on the product of the regulator gene.

How the unnecessary, or inactive, genes in a fully differentiated cell are inactivated has not been fully explained but it is widely believed that a primary protein, histone, is in some way combined or associated with these genes to inhibit them from further activity. Just how permanent or irreversible the inhibition of these genes may be will be discussed further under the heading 'Regeneration and repair'.

Other common examples of induction are the induction of the lens from the overlying ectoderm by the optic cup and the induction of the kidney tubules (nephron) from the metanephrogenic cap by the terminal branches of the ureteric bud. Induction is such a frequent phenomenon that development could be regarded as a branching and ever-widening series of inductive processes with the induction of the neural plate by the notochord as the initial or at least a very early step—hence the name *organizer* or *organization centre* given to the dorsal lip of the blastopore where the notochordal cells lie. Nor does induction consist entirely of one-way traffic; in many cases, e.g. in the development of the limb bud, there appears to be a series of interactions between the two tissues, ectoderm and mesoderm, and it is extremely difficult to determine which of these was responsible for initiating the processes of development and differentiation.

NUCLEAR TRANSPLANTATION

Associated with the subject of differentiation is the question of whether the differentiated cell actually retains all the genes it originally possessed. The apparent irreversibility of differentiation suggests that the unused part of the genome is lost but there is no real evidence for this view. Indeed, from recent work with nuclear transplantations, there is now more evidence for the belief that the genome is intact. These investigations (Fig. 11.3) began with the

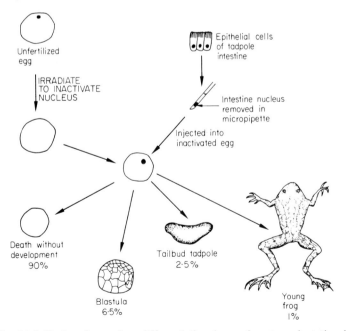

Fig. 11.3 Testing for nuclear differentiation by nuclear transplantation in amphibia. The numbers represent the proportions of eggs with transplatned nuclei which survive to the particular stages. From Truman, D.E.S. *The Biochemistry of Cytodifferentiation.* Blackwell Scientific Publications, Oxford.

separation or disaggregation of cells of the early amphibian embryo at the blastula, gastrula and neurula stages. One of these cells was sucked into a micropipette narrow enough to break the cell without damaging the nucleus. The contents of the cell, including the nucleus, were then injected into an unfertilized ovum after its nucleus had been removed or killed. Similar experiments have also been done with nuclei from cells lining the intestine of tadpoles. Only a relatively small number of such 'manufactured' ova succeeded in developing into normal adults but they are evidence that nuclei of differentiated cells do retain all their genes and that the apparent loss is due mainly to the difficulty of unmasking the genes.

DEDIFFERENTIATION

When cells from an embryo or adult are grown *in vitro* the behaviour of the cells often suggests dedifferentiation. They lose many of those features which characterize their normal adult state and assume a relatively undifferentiated appearance. For instance, mesenchymal derivatives such as loose connective and muscle tissue cells 'dedifferentiate' into apparently identical primitive mesenchymal cells *in vitro* but are still distinguishable from different epithelial and glandular cells grown *in vitro* although these too are very similar to one another under these conditions. This is a regular and apparently essential step in their multiplication and will be discussed later with regeneration and repair as well as with the subject of cell culture.

HORMONES AND DIFFERENTIATION

It should not be imagined that differentiation of a cell depends entirely on one inducing substance which, whenever it impinges on a cell, triggers off a series of reactions within it leading to the differentiated state. In the first place the cell must be receptive to the action of the particular inducer i.e. the cell must have progressed the required distance along its developmental path and the conditions of gene activity or inactivity as well as metabolic reactions must be just right. Equally important, if the arrival of the inducer is late it will probably be quite ineffective because conditions within the cell have moved beyond the stage when that particular inducer could have an effect. These changes and the progression up to and beyond the stage when one inducing substance could act suggests, probably correctly, that a series of inducing substances or outside influences is required for the differentiation of a cell—many of them ill understood or even unknown to us but they are not haphazard in their timing. Examples of inducing substances not usually regarded as such but obviously as significant as any classical inducer are the endocrine secretions, e.g. thyroxin, which in animals reach the cells via the blood stream from a distance but can nevertheless stimulate or alter differentiation in a distinctive fashion. In human development, hormonal effects are best exemplified by the action of the gonado-trophic hormones on the reproductive system at puberty and, at approximately the same time, the action of growth hormone on the epiphyseal cartilaginous plate in long bones.

In plants this action of hormones is also well known and diffusion of the hormones plays an all-important part in the growth and differentiation of tissues, the different hormones acting in concert to produce different effects. The action of these hormones may also vary according to the concentration of the hormone, the location of the target cells and to the stage of development of the plant.

FURTHER READING

Ashworth J.M. (1973) *Cell Differentiation.* Outline Studies in Biology. London: Chapman & Hall.

Beerman W. *et al.* (1969) *Cell Differentiation and Morphogenesis.* Amsterdam: North-Holland Publishing Company.

Bellairs R. (1971) *Developmental Processes in Higher Vertebrates.* London: Logos Press Limited.

Berrill N.J. (1961) *Growth, Development and Pattern.* San Francisco & London: W.H. Freeman and Company.

Berrill N.J. (1971) *Developmental Biology.* New York & London: McGraw-Hill Book Company.

Brachet J. (1974) *Introduction to Molecular Embryology.* London: The English Universities Press Ltd.

Butler J.A.V. (1968) *Gene Control in the Living Cell.* London: George Allen and Unwin Ltd.

Gurdon J.B. (1973) *Gene Expression during Cell Differentiation.* Oxford Biology Readers. Oxford: University Press.

Hamburgh M. (1971) *Theories of Differentiation.* London: Edward Arnold.

Markert C.L. & Ursprung H. (1971) *Developmental Genetics.* Englewood Cliffs, N.J.: Prentice-Hall Inc.

Oppenheimer J.M. (1967) *Essays in the History of Embryology and Biology.* Cambridge, Mass.: The M.I.T. Press.

Saxen L. & Toivonen S. (1962) *Primary Embryonic Induction.* London: Logos Press Limited.

Spratt N.T. Jr. (1971) *Developmental Biology.* Belmont California: Wadsworth Publishing Company, Inc.

Sussman M. (1973) *Developmental Biology, Its Cellular and Molecular Found-ations.* Englewood Cliffs, N.J.: Prentice-Hall, Inc.

The Open University (1973) *Genes and Development* Units 1–3, 4–6. Milton Keynes: The Open University Press.

Truman D.E.S. (1974) *The Biochemistry of Cytodifferentiation.* Oxford: Blackwell Scientific Publications.

Waddington C.H. (1966) *Principles of Development and Differentiation.* London: Collier-Macmillan Limited.

Willier B.H. & Oppenheimer J.M. (1964) *Foundations of Experimental Embryology.* Englewood Cliffs, N.J.: Prentice-Hall, Inc.

12 REGENERATION AND REPAIR

Although regeneration is still a poorly understood phenomenon, it is an intriguing one and its further investigation is sure to yield information which will help to interpret other phenomena in the field of development.

The marked difference in regenerative capacity between the invertebrates and the vertebrates is well known, the former showing so much regenerative power that some worms, for instance, can be cut in two and thereafter be able to restore all the missing parts in each half; at the other end of the scale, i.e. in mammals, regeneration is at a minimum and so unlike what is normally considered as regeneration that it is simply called repair. It is not enough to say that invertebrate cells are simpler or less complex than vertebrate cells or that regenerative capacity diminishes with evolutionary progress; the aim must be to discover the molecular or other differences which account for the gross functional contrasts but we are still far from understanding or even recognising them.

LIMB REGENERATION

In an amphibian larva such as the newt, regeneration can be studied closely and a great deal has been gleaned from the experiments already done on this animal. When a newt's limb is amputated, the wound closes quickly by overgrowth of the epithelium but this is accompanied by a subepithelial proliferation of cells forming the stump blastema. These cells appear to be undifferentiated or at least poorly differentiated, showing no evidence at first that they are derived from or intend to become the normal adult tissues of a limb, e.g. muscle, bone, cartilage or connective tissue. Yet these cells do eventually differentiate to form adult tissues, organizing themselves spatially and quantitatively to reproduce a normal limb.

Origin of blastema cells

This phenomenon presents many problems; what, for example is the origin of the cells comprising the stump blastema? One view considers that they are mobilized from the rest of the animal's body, migrating to the stump and differentiating to form adult tissues. Amongst invertebrates there is some support for this theory because uncommitted undifferentiated cells called neoblasts have been recognized scattered throughout the tissues and they could be the source of the stump blastema cells. An alternative view is that the blastema cells arise by 'dedifferentiation' of the cells left in the stump after amputation and evidence for these being the source comes from an experiment in which amputation was followed immediately by the grafting of a limb from another amphibian larva which had recognizable differences in its nuclei compared with those of the host animal. After the grafted limb had become established, it in turn was amputated, care being taken to leave a thickness of tissue from the grafted limb in the stump. Subsequent regeneration of the limb was identified, by histological examination, as arising from the cells of the grafted limb, an indication that mobilization of cells from the remainder of the animal is unnecessary and unlikely. Further evidence for this view is obtained by irradiating the stump immediately after amputation or by irradiating the rest of the animal's body (excluding the stump); in the former case, no regeneration occurs and in the latter the limb is regenerated normally.

Arising from these observations is the question of how much dedifferentiation occurs in these stump cells, i.e. do they dedifferentiate to such a degree that, after proliferation, they can differentiate along a pathway quite different from that which they took during the normal development of the limbs. In simple terms, can the residual muscle tissues, for instance, dedifferentiate to such an extent that its cells may later differentiate into muscle or cartilage or connective tissue? It is an attractive hypothesis in view of the featureless appearance of the blastema cells but, after all, the histological characters and proliferative capacity of the cells do not necessarily imply that nuclear i.e. chromosomal, dedifferentiation has occurred. Opinion, at present, favours retention of chromosomal differentiation with only a simulated dedifferentiation, associated with and perhaps necessary for the multiplication of numbers.

Regeneration and nerve supply

There is still the fundamental problem of why the limb of the newt can regenerate and the limbs of the bird or mammal cannot. As well as trying to discover the answer for the sake of science, the information may have far-reaching practical applications. There are no apparent histological differences in muscle, bone, cartilage or connective tissue between newt and mammal. One feature of possible significance has been noted and followed up, namely the nerve supply to the limb. Regeneration in a limb does not involve nerve cell bodies because there are few of these in a limb but regeneration of nerve cell processes (or fibres) is certainly a factor; in all animals, nerve cell processes do grow out again from the proximal cut ends although the normal pathway and distribution may be disturbed. In an amphibian larval limb, the cross-sectional area occupied by nerves is far greater relative to the total cross-sectional area of the limb itself compared with that in an animal which cannot regenerate a limb. This feature prompts the suggestion that the amount of nerve in a limb plays a part in determining whether regeneration will occur. Evidence in favour of this view is found when the ends of the nerve fibres are turned away from the site of the limb at the time of amputation; the limb under these circumstances fails to regenerate. If, however, the neural folds in the embryo are removed and the limb, without innervation, i.e. aneurogenic, is allowed to develop, then subsequent amputation is followed by regeneration, a phenomenon which calls for further investigation of the whole problem of limb regeneration relative to the nerve supply.

As well as the features mentioned above, there is of course the more obscure problem of how the cells of the stump blastema not only differentiate into tissue but can also organize and reproduce the normal arrangement of muscles, bones and tendons etc. in the new limb. We know very little about the mechanisms or factors involved in producing a new limb (or for that matter the development of a normal limb) but if further examination of the phenomenon of regeneration can yield new facts the whole field of pattern formation in growth will benefit.

Tail regeneration

At the evolutionary stages when limb regeneration is still possible, other parts of the body also have considerable regenerative capacity. The tail, for instance, can be regenerated by larval anurans and urodeles, the presence of the notochord in the stump being required for complete regrowth. Tail regeneration is also a common feature in the lizard but not in other reptiles. For the lizard, which can readily discard its tail when caught in this way, the ability to regenerate that part is accompanied by special structural features in its tail. An important but negative feature of tail regeneration is that it is not the retention of an embryonic feature. Indeed, embryonic lizards cannot regenerate amputated tails.

REGENERATION AND REPAIR IN MAMMALS

Wound healing

Higher vertebrates are not entirely bereft of the ability to regenerate their tissues or even organs but the capacity is more restricted and might better be considered as a process of repair. In a wound of the skin, for example, the blood clot filling the gap after bleeding stops is soon covered by epithelium which grows from the edges of the wound. Soon afterwards the cells i.e. fibroblasts, of the dermis also proliferate and invade the blood clot along with new blood vessels. As well as removal of the remains of the clot by macrophages which accompany them, the fibroblasts soon bridge the wound and form a connective tissue which becomes progressively firmer until only a dense white scar is left. The fact that repair of a skin wound is achieved by the formation of this scar and not by normal dermis (like that of adjacent skin) implies that the repair process is not perfect but there is little doubt that basically the stimulation and proliferation of the cells in adjacent tissues represent a form of regeneration. Another example of repair is illustrated by the healing of fractures in bone where the cells which react by producing new bone are derived from the periosteum of the old bone near the broken ends.

Organ regeneration

If proliferation of the remaining cells after removal or destruction of some of the original population can be considered as regeneration, then there are many examples even in the mammal and human. If a large amount of liver is removed and the animal allowed to recover from the operation, examination of that organ some weeks later reveals that the liver has regained most if not all of its volume by proliferation of the remaining cells. The same applies in the case of other glands which are partially removed and even to the kidney, after all but a part of one kidney is resected. However, kidney must behave in a rather different way when it regenerates. In the liver, where the individual liver cell is the functional unit, functional capacity relative to the general metabolic processes of the body can be restored by multiplication of any of the liver cells. In the kidney, however, the unit of function as far as the well-being of the body is concerned is the nephron and a complicated unit like that cannot be replicated. The increase, therefore, in the size of the kidney following removal of most of that tissue consists of multiplication of the cells in the Bowman's capsules and tubules giving bigger nephrons, e.g. longer tubules, with enhanced ability to filter and reabsorb.

In the repair of a skin wound, in the healing of a fracture or even in regeneration of the newt limb, the stimulus and subsequent reaction i.e. cellular proliferation, are essentially local and there need be little or no involvement of the rest of the body. We could visualize the loss, after wounding, of a chemical substance or secretion which is normally produced by the cells of a tissue and normally maintained at a high enough local concentration to inhibit mitotic division of the same cells. Tissue specific substances (proteins) called chalones have been identified as fulfilling this function in many tissues and their loss or leakage from tissue at the site of a wound may be the initial stimulus for the reparative or regenerative process. This hypothesis does not explain the selective and accurate differentiation of the cells once they have proliferated. In the case of liver and kidney regeneration however, other factors are probably also involved; experiments have shown that the stimulus to regenerate may also be related to failure of the remaining cells or tissue to deal adequately with its normal *functional load*. From a variety of experiments, there is evidence of a humoral factor stimulating regeneration, e.g. an abnormal substance circulating in the blood as a result of faulty or inadequate metabolism by the remaining cells or tissue. This is an attractive hypothesis because it may also be used to explain the precise control of organ size during normal growth, i.e. when an organ like liver or kidney is large enough to deal with a particular metabolic process in the body, it is no longer under stimulation to grow from some by-product of incomplete metabolism.

REGENERATION IN PLANTS

It would be quite wrong to think that regeneration is limited to the features described so far. Certainly in many invertebrates the degree of regeneration is extensive and surprising in its complexity, e.g. regeneration of the entire head region after amputation, but an excellent and far more common example of regeneration occurs in plants. 'Cuttings', i.e. a part of the stem with perhaps a few leaves may be taken from many kinds of plants and, when planted in soil, proceed to form a root system complete with the normal tissue architecture and function. Perhaps the cellular proliferation in these cases is less surprising than the differentiation of completely different tissues which develop from the original cutting. Some plants may even be grown *in toto* from a leaf or even part of a leaf and we begin to suspect a more advanced degree of dedifferentiation of these proliferating cells than normally occurs in mammals (see also Cell Culture).

FURTHER READING

Berrill N.J. (1961) *Growth, Development and Pattern.* San Francisco & London: W.H. Freeman and Company.
Goss R.J. (1969) *Principles of Regeneration.* New York: Academic Press.
Goss R.J. (1972) *Regulation of Organ and Tissue Growth.* New York: Academic Press.
Kiortsis V. & Trampusch H.A.L. (1965) *Regeneration in Animals and Related Problems.* Amsterdam: North-Holland Publishing Company.
Kuhn A. (1971) *Lectures on Developmental Physiology.* Berlin & New York: Springer-Verlag.
Saunders J.W. Jr. (1970) *Patterns and Principles of Animal Development.* London: Collier-Macmillan Limited.

13 TWINS AND MULTIPLE BIRTHS

Twins intrigue everyone, although for different reasons; and twins are of two kinds, the *monovular* or identical and the *binovular* or non-identical. Compared with its frequency in most other animals, the binovular type is unusual in the human; multiple births or litters which are very common in the animal world are the result of several ova being released simultaneously then fertilized and implanted together. Many factors determine the usual number of fetuses in a pregnancy, e.g. size of fetus, size of uterus, length of gestation, the frequency and other features of the oestrous cycle, the degree of maturation at birth, etc.—a mixed bag of reasons mostly associated with evolution, environment and the need for maternal care of the young. The important point is that in the human and some others, such factors have limited the human to single births except in odd infrequent cases where two or more fertilized ova implant in a uterus 'designed' to carry only one. These fetuses, having implanted at separate sites, have separate fetal circulations and separate membranes, with limited accommodation as their most serious problem.

During gestation in an animal in which twins are uncommon their presence is an embarrassment to the uterus and indeed to the whole maternal organism. At delivery which is often earlier than usual, mechanical problems are a hazard but lessened by the fact that the twins are as a rule smaller than a singleton; the mother bearing twins often shows greater metabolic and hormonal disturbances than normal.

How do monovular twins arise?

Twins and other multiple births of the non-identical type in the human are like brothers and sisters in the same family and as different, whereas monovular twins, because they are derived from the same ovum, have identical genotypes, are of the same sex and look like one another; they are unique and always in demand as interesting phenomena or objects for study. One problem, for instance, concerns the time at which monovular twins arise and apparently twinning can occur at more than one developmental stage (Fig. 13.1); theoretically at least, the end of the first cleavage may be the time when the two cells dissociate and proceed to form two separate morulae and later two blastocysts, but complete separation at the first cleavage would require the disappearance of the zona pellucida at that time, an unlikely but not impossible feature. If this were

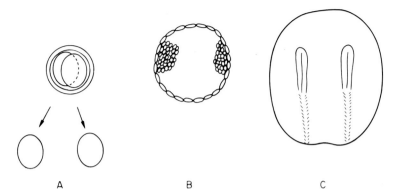

Fig. 13.1 Monovular twinning; (A) by separation of the two cells after the first cleavage; (B) by the formation of two inner cell masses within the blastocyst; (C) by the development of two embryonic axes on the same disc.

the case, however, complete separation of the two embryos would give two separate fetuses each with its own extra-embryonic accessories, as in binovular twins but of course with identical genotypes. Closely sited implantations may lead to fusion of the two fetal circulations in a common placenta.

At the blastocyst stage, two clearly defined inner cell masses may develop within the trophoblast shell and each is then capable of becoming an embryonic disc with its own amniotic cavity. Further development gives rise to twins each with its own amnion but enclosed in a common chorion; two umbilical cords are present leading either to separate placentae or to one large placenta in which the circulations may intermingle freely.

The third and latest stage at which identical twins may develop is in the very early embryonic disc. Here the process of twinning begins with the appearance of two organization centres which produce two embryonic axes in the disc and later two fetuses within the same amnion and chorion. One or two umbilical cords and one fused or two separate placentae are also present. All will go well with the two (or perhaps more) embryos developing from the same disc provided that the embryonic axis are far enough apart to allow them to acquire their own head and lateral folds for the primitive body form stage. If not, and depending on the exact relationship of the axes to one another, all sorts of strange *double monsters* or *Siamese twins* (the commonest being thoracopagus twins) will develop. Fig. 13.2 shows how some of these monsters arise. A study of these monsters reveals interesting anomalies in organs such as heart and intestine occurring as a result of partial fusion.

During the development of identical twins with fusion of the circulations, it is conceivable that imbalance in the strength of the hearts and/or positioning of the umbilical cords on the placenta may lead to weakening or even death *in utero* of one twin but in the great majority of cases identical twins are of the same weight and size at birth.

There is no doubt that identical twins are of interest in many respects. Even

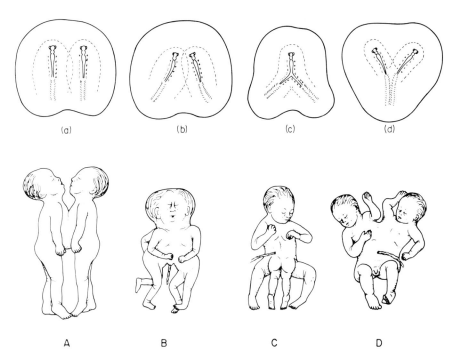

Fig. 13.2 (A), (B), (C) and (D); four types of conjoint twins with the corresponding orientations (a, b, c and d) of their embryonic axes on the discs. (A) from Hamilton, W.J. and Mossman, H.W., *Human Embryology*, 4th edition. Wm. Heffer & Sons Ltd., Cambridge. (C) and (D) from Patten, B.M., *Human Embryology*, 3rd edition. McGraw-Hill Book Company, New York, London.

in their inception there is some mystery and the departure from normal development to give duplication, e.g. of the inner cell mass or of the organization centre, suggests abnormality. Whether it is purely accidental (which is an admission of our ignorance) or whether teratogenic factors are involved is not clear but probably the latter explanation is the more likely. Amongst lower animals, birds and fishes, twinning of the identical type occurs following sharp changes in temperature during the early stages of development, i.e. at or before the appearance of the organizer. In birds and fishes which have large amounts of yolk in their ova and hence develop with yolk sacs, the presence of identical twin embryos on the same yolk sac leads to insurmountable problems at the stage when the yolk sac is due to be absorbed into the organism.

Monovular or binovular?

After birth, twins are often investigated to determine whether they are identical or not. Although monovular twins should be identical in every physical feature since they have identical genotypes, exceptions do occur because, even when they develop in the same uterus, chorion and amnion, minor differences in the environment may cause structural and physiological discrepancies in development. An examination of the fetal membranes (amnion and chorion) and the placental circulations can give some indication of the type of twinning which has occurred. Also useful is the degree of similarity in the facial features (colour of hair and eyes, shape of nose, ears, mouth etc.); comparison of the finger, palm and sole prints has been used as well, reliance being placed on the degree of similarity because not even proven monovular twins have identical patterns; data on the blood groups of the twins may help and, with the introduction of the more elaborate techniques in this field, may prove even more reliable in the future. There is however one method of identifying identical twins with assurance, namely by the exchange of skin grafts between them, retention of the grafts indicating that they are genetically identical. In the human this method is almost infallible but not absolutely so because it is just possible that fusion of the two circulations belonging to binovular human twins may have occurred early in gestation, permitted the exchange of blood cells between the fetuses and made them tolerant to one another's tissues. This often happens in twin cattle pregnancies (see freemartins—Immunological Aspects).

The effect of the post-natal and childhood environment, whether it be nutritional, educational or social, on the development of an individual can be conveniently studied in identical twins who have been brought up separately from birth onwards, and a great deal of useful information has been obtained. The records of such investigations make interesting reading and the findings indicate that environment has a powerful influence on development.

Multiple births in the human, e.g. triplets, quadruplets and quintuplets are rare and the diagnosis of how they arose (an academic exercise, after all) may be difficult, depending as it does on the information available at the time. In the case of the famous Dionne quintuplets, the details of the placenta and its membranes were meagre—one large placenta, one large chorion and probably five amnions and five umbilical cords; but the quins were all of the same sex and same blood group and sufficiently similar in facial features and finger, palm and sole prints to lead to the assumption that they were all derived from the same ovum.

The spate of multiple births recorded recently is associated with the administration of 'fertility' drugs, i.e. gonadotrophins like the follicle-stimulating hormone. These are substituted for the normal stimulus which was judged to be absent or too small to be effective in the development, maturation and rupture of the Graafian follicle with release of the ovum. The difficulties of accurately estimating the deficiency and the dangers of inaccurate evaluation are illustrated by the occasional unexpected result, although these should not overshadow the less well publicized successes.

FURTHER READING

Bulmer M.G. (1970) *The Biology of Twinning in Man.* Oxford: The Clarendon Press.

Hamilton W.J. & Mossman H.W. (1972) *Human Embryology* 4th ed. Cambridge: Wm. Heffer & Sons Ltd.

Harrison G.A., Weiner J.S., Tanner J.M. & Barnicot N.A. (1964) *Human Biology.* Oxford: The Clarendon Press.

Medawar P.B. (1957) *The Uniqueness of the Individual.* London: Methuen.

Osborne R.H. & De George F.V. (1959) *Genetic Basis of Morphological Variation.* Cambridge, Mass.: Harvard University Press.

Tanner J.M. (1962) *Growth at Adolescence* 2nd ed. Oxford: Blackwell Scientific Publications.

14 IMMUNOLOGICAL ASPECTS OF DEVELOPMENT

The immune response in adult organisms is well known, and only its salient features will be mentioned here. A piece of tissue or an organ transplanted from one individual to another is normally rejected or discarded by the host. Skin, for example, transplanted in this way appears to 'take', i.e. it becomes warm and vascular within a few days and remains so for perhaps ten days; then, even when the best surgical procedures have been used, it becomes cold and white and eventually dies; in the end it is discarded as a slough. A second attempt to transplant skin (or any other tissue) from the same donor to the same host meets with an even more rapid rejection, e.g. in 7–8 days, and the same response occurs with any subsequent attempt. The explanation, briefly, is as follows: after the first transplantation, macromolecular protein particles, *antigens*, from the grafted tissues are picked up by the lymphatic vessels at the site of the transplant and carried to the lymph nodes draining that region, e.g. in the arm all the lymph vessels converge on the axilla and their contents are filtered through the lymph nodes there. The particles derived from the graft are probably phagocytosed by macrophages in the lymph nodes and in the same way information about them is passed to the adjacent proliferating lymphocytes. The lymphocytes react by producing substances which can neutralize or destroy these particular particles or antigens. At the molecular level, the process comprises recognition of these particles as foreign to the body, followed by the stimulation of part of the genome in the lymphocyte and, through the transcription, translation and protein-synthesizing mechanisms, to produce globulins known as *antibodies* and capable of attacking the antigens which induced their formation. Lymphocytes thus stimulated are released from the lymph nodes, carried along the lymph channels to reach the blood stream and disseminated throughout the body via the blood vessels. At the site of the graft, the lymphocytes aggregate, escape from the capillaries and collect in the tissues, reaching close up to the graft. At this distance the antibodies seem to be able to destroy the cells producing the foreign protein to which the lymphocytes were sensitized. The more rapid response to a second attempt at grafting skin or tissue from the same donor occurs because the process of sensitization of the lymphocytes does not have to be repeated. Sensitized lymphocytes or their percursors will remain in the lymph nodes long after the initial response, still capable of mobilization and proliferation for defence against another 'invasion' by tissue from that particular donor. Notice that the reaction is donor specific, not tissue specific. Numerous attempts, some very successful, have been made to subdue this natural response to the introduction of tissue from another individual but these do not concern us here.

The reactions of the body (a) to the implantation of 'foreign' tissue and (b) to invasion by micro-organisms, e.g. bacteria, are basically the same, viz. the production of specific antibodies to neutralize or destroy the foreign proteins but there are important differences. For instance, the reaction to implanted tissues (cell-mediated response) involves bringing the cells close to the foreign material because, apparently, these antibodies can operate only by cell contact or over very short distances; on the other hand the reaction to invading organisms, which can disseminate widely in the body, requires a high concentration of the antibody in the blood stream, the so-called *humoral response*.

IMMUNOLOGICAL TOLERANCE

In this text, we must refer to three 'naturally' occurring situations in which the exchange of grafts between individuals is possible. The first, probably of less significance to developmental biologists, concerns grafts between *genetically identical individuals*. Identical twins are the best example because they are developed from the same ovum but homeogenetic stocks of animals, e.g. inbred mice or rats, have been developed and are useful for experiments investigating the phenomena of the immune response. It appears that complete identity of the whole genome is not essential for successful grafts and indeed a limited number of gene loci have been identified as responsible for the immune response; some loci or genes are more powerfully antigenic than others and indeed the search for human graft donors and hosts usually aims at approximation rather than complete compatibility of the genomes, the possibility of success (aided by drug protection) deriving from compatibility of the more strongly antigenic genes.

The second natural situation occurs when *fetal* or *early post-natal hosts* are used. Before hatching and for about ten days thereafter, the chick will accept grafts from even an adult donor and retain them. Mice and rats behave in the same way during fetal life and for about 10 days after birth but not all animals show this immunological tolerance (absence of response) so late in their life cycle. Many animals, e.g. guinea-pig and human, can show the response long before birth. When we consider that the whole immune response and immunological competence is dependent on genetic factors it may at first be surprising that the embryo does not react from the earliest moment of its existence. Immunological competence, however, has to develop in the individual or more correctly the mechanism whereby it can be effected has to develop. Birth is, of course, no criterion of the stage of development and we have to correlate the emergence of the immune response with the development of some tissue or system in the body. The first clue to the nature and *ontogeny of the immune response* came with the discovery that removal of the thymus gland (thymectomy) in a mouse, before it had achieved immunological competence, prevented it from rejecting any grafted tissues thereafter. It was also apparent that such an animal, known as a *runt*, succumbed to infections which the normal litter mate could survive; in other words, the development of all defence mechanisms was inhibited. Thymectomy *after* the development of immunological competence has no inhibitory effect and it was deduced that, while the thymus is responsible for establishing immunological competence, it apparently plays no part in maintaining it. It should be noted that the natural deterioration and diminution in size of the thymus gland after puberty in the human occurs without significant loss of immunological competence.

It is not yet absolutely clear how the thymus *functions* but the results of some experiments help us to understand the mechanism involved. For instance, thymectomy before the establishment of the immunological competence does not prevent lymphocyte and lymphoid tissue development although the amounts of these are reduced. Replacement of the thymus gland after thymectomy in the incompetent animal will establish normal immunological competence as will the injection of immunologically competent cells (lymphocytes) from a mature individual genetically identical to the host. Replacement in the animal of thymic tissue enclosed in an envelope impermeable to cells can also create immunological competence and increase the amount of lymphoid tissue. The fact that tolerance to a tissue grafted before the development of any immune response is followed by tolerance to all tissues from that same donor after 'natural' competence develops suggests that, in some way, the development of immunological competence is associated with the identification as 'self' of all tissues present in the body at that time whether or not they are genetically identical to this host. Is there a kind of inventory drawn up at that time, of all the tissues in the body, with anything foreign which appears subsequently being naturally rejected? This is true even when the immunological defences of the body are confronted by tissues or proteins which, although genetically identical to the host and actually present in the body, were hidden from immunological detection. They were not included therefore in the 'inventory' prepared when immunological competence was established, but may evoke a response and be rejected or destroyed later. Proteins of this nature are sometimes known as 'sequestered' antigens and the best examples are the spermatozoa and the colloid in the vesicles of the thyroid gland. The former which are not only differentiated relatively late in life but are also shielded by the wall of the seminiferous tubule from immunological detection, can, when injected into the tissues of the animal, initiate an immune response to their own proteins and inhibit further development of spermatozoa. A condition known as thyroiditis is thought to be an autoimmune reaction arising in the following way: the colloid in the thyroid vesicles is a sequestered antigen because it has remained out of reach or hidden from immunological detection during life but, as a result of trauma or some other cause, escapes into the tissues and initiates an immunological reaction: sensitized lymphocytes then invade the gland and attack the cells which produce the colloid. Some immunologists however are sceptical of such a simple explanation of these autoimmune reactions.

An intriguing situation may develop in binovular twins because of the period of tolerance prior to the development of immunological competence. If the placental circulations of these twins should become linked temporarily or permanently in the pre-competent phase, there will be an exchange of tissues, e.g. red and white blood cells, which are then accepted as 'self' by the two fetuses; from then onwards each fetus will also accept grafts of any tissue from the other and as far as this test is concerned they will behave as identical (monovular) twins. Fusion of the circulations occurs only rarely in human binovular twins but is a regular feature with cattle twins. In the latter, however, the immunological aspects were noted only comparatively recently, yet the intermingling of the fetal circulations has long been recognized as responsible for the abnormalities in the genitalia of a female calf when the other twin is a male. It is generally believed that the early development of the male genitalia (which normally occurs before that of the female genitalia) produces sufficient male sex hormones to diffuse into the circulation of the female calf before its genitalia have developed and to direct their development towards maleness. The result is a *freemartin* or a sterile female calf.

MATERNAL TOLERANCE OF THE CONCEPTUS

The third and most surprising 'natural' type of immunological tolerance is that of the mother to the fetus *in utero*. In the case of central implantation with an intact uterine epithelium throughout gestation (epitheliochorial placenta), this is understandable but when implantation is interstitial with a haemochorial type of placenta it is difficult to understand how the fetus can be tolerated throughout a long pregnancy without signs of rejection developing or without evidence of maternal antibodies to the fetal tissues which, of course, contain paternal genes.

There are several possible explanations of the phenomenon: for instance, since steroids, e.g. cortisone, can reduce or prevent rejection responses such as lymphocyte proliferation after tissue transplantation experiments and operations, it has been suggested that, during pregnancy, the mother develops a natural tolerance to foreign tissue because of the increased secretion and concentration of steroids in her body. This cannot be the whole explanation because a pregnant mother can still reject ordinary skin grafts although the response may be somewhat delayed.

The second possibility is that the uterus, on account of the endometrium having such a poor lymphatic drainage, is a privileged site for implantation i.e. foreign tissue embedded in it would not be easily recognized immunologically;

experimental work involving tissue implants in the uterus lends some support for this view.

More interest has been focused on the trophoblast as an effective barrier to immunological detection of fetal tissues. The cells of the trophoblast normally form a complete shell for the fetus during gestation (thus enclosing the *conceptus*) and is in an ideal position to behave as an immunological shield. It has been suggested that the trophoblast cell does not emit antigens and is therefore not recognized as a foreign implant by the mother. On the other hand, it could be that the trophoblast cell *is* antigenic and that a thin layer of fibrinoid material, which has been demonstrated by the E.M. lying between the trophoblast and maternal tissue, forms an impermeable barrier to the passage of antigens from the fetus. A great deal of work has still to be done in this field before maternal tolerance to the fetus is fully understood. Although the transfer of antigens between fetus and mother is very effectively blocked under normal circumstances, the passage of antibodies across the placental barrier occurs far more easily particularly if the placenta is of the haemochorial type (see below).

A well-known example of breakdown in maternal tolerance occurs in association with the rhesus (Rh) factor which comprises a group of antigens present on or in the human red blood cell. About 85 per cent of the population carry these antigens and are called Rh positive while the remainder are Rh negative. If an Rh-ve mother bears a child which has inherited the rhesus antigens from its father, there is theoretically little chance of the mother's tissues recognizing the presence of the factor in the fetus but if an escape of fetal red blood cells does occur into the maternal circulation, the mother will develop antibodies to the RH + ve blood. Apparently this can and does happen more frequently than we would expect, particularly at the end of pregnancy or during delivery of the child when traumatic damage to the placental barrier is likely. Typically, the mother is sensitized to the Rh antigens only at this stage of pregnancy, too late perhaps to form antibodies to that child but, in a subsequent pregnancy, also with an Rh positive baby, the Rh antibodies attack and destroy (haemolyze) the fetal red blood cells before the end of pregnancy.

The free passage of antibodies across the placenta has an unfortunate result in such a case but in other circumstances the newborn child benefits from the transference of antibodies from its mother. For instance, in the human, maternal antibodies formed in response to infectious diseases before pregnancy can cross the placenta to reach the fetal blood stream and, if in sufficient concentration, can provide a passive immunity for a short time after birth to the same infectious diseases. In animals with a thicker morphological placental barrier, antibodies do not pass across freely and the newborn has to rely for its passive immunity to infections on the antibodies coming over in the milk and being absorbed through the wall of the gut.

Graft-versus-host reaction

Finally, and briefly, there is the graft-versus-host reaction, which is an experimental phenomenon scarcely if ever occurring naturally. The reaction occurs when fully competent lymphocytes from an adult animal are injected into an immunologically tolerant host, e.g. a fetus in the pre-competent phase. The host cannot recognize and react against foreign material but the injected lymphocytes can; they become sensitized to the host tissues, proliferate freely in its lymphoid tissues and damage them severely. The host then becomes a runt, incapable of defending itself against 'normal' infections. Not only can the reaction be demonstrated after the injection of immunologically competent lymphocytes into a newborn rat or mouse (it is not yet competent to deal with foreign material) but the response can also be shown to occur in an adult animal X-irradiated to destroy its immunological defences and thereafter injected with lymphocytes from another (genetically different) animal. A more intriguing experiment consists of mating one animal from a genetically homogenous stock (AA) with one from another genetically homogenous stock (BB) to obtain hybrids (AB). If an adult hybrid is injected with lymphocytes from one parent, e.g. BB, the injected cells are 'accepted' by the hybrid because it has the 'B element' in its own genotype and recognizes it as 'self' but the injected cells recognize the 'A element' in the hybrid's genotype as foreign and reacts against its tissues giving a graft-v-host reaction.

FURTHER READING

Assali N.S. (1968) *Biology of Gestation* Vol. 2 Chapter 7. New York & London: Academic Press.

Barnes A.C. (1968) *Intra-uterine Development* Chapters 17 & 24. Philadelphia: Lea & Febiger.

Ebert J.D. & Sussex I.M. (1970) *Interacting Systems in Development* 2nd ed. Chapter 5. New York & London: Holt, Rinehart & Winston, Inc.

Medawar P.B. (1957) *The Uniqueness of the Individual.* London: Methuen.

Roitt I.M. (1971) *Essential Immunology.* Oxford: Blackwell Scientific Publications.

Solomon J.B. (1971) *Foetal and Neonatal Immunology.* Amsterdam: North-Holland Publishing Company.

Turk J.L. (1969) *Immunology in Clinical Medicine.* London: Wm. Heinemann Medical Books Limited.

Weber R. (1965) *The Biochemistry of Animal Development* Vol. 1 Chapter 7. New York & London: Academic Press.

Weir D.M. (1970) *Immunology for Undergraduates.* Edinburgh & London: E. & S. Livingstone.

15 CONGENITAL ABNORMALITIES AND TERATOLOGY

Every child born with a congenital abnormality, i.e. abnormal development of some part of its body or of some function or ability, represents a tragedy. The defect in development occurs such a long time before it is discovered that at birth little can be done apart from reparative or palliative measures. There are no second chances in the course of fetal development and furthermore, the cause of the abnormality is often extremely difficult or impossible to ascertain. With up to 5 per cent of children suffering in this way, the problem is a serious one because they require treatment and extra care to keep them alive and perhaps for the rest of their lives, particularly when mental retardation is also present.

It is a relatively simple task to catalogue the innumerable factors which may give rise to congenital abnormalities, and *teratology*, or the study of these factors and how they act, is a serious and rewarding undertaking from many points of view. All the different causes will not be listed here but the broad groups into which they fall will be discussed. Anything which interferes with the normal process of development must be considered and, in general terms, we can accept that they form two main groups: (a) abnormalities inherited from the parent(s) through the chromosomes or genes and (b) environmental factors which can adversely influence the otherwise normal process(es) of development.

CHROMOSOMAL AND GENETIC ABNORMALITIES

The correct number of chromosomes is essential for normal development although experimentally triploidy (3n) tetraploidy (4n) and other multiplications of the haploid (n) state do not cause as much variation in general or regional development as might be expected. Indeed it appears fundamentally more important for normal development to have the balanced or co-ordinated activities of the complete set (2n) or sets (3n, 4n) of chromosomes than to have a variation (aneuploidy) e.g. $+1$ or -1 etc., in the normal complement. Such loss of balance in chromosomal activity arising from the absence of or extra chromosomes is sometimes blamed for the numerous early embryonic deaths— abortions or resorptions—which are otherwise inexplicable. An increase in the normal number (two) to three of the homologous chromosomes (giving a

162

trisomy) or the loss of one chromosomes has now been shown to be responsible for some abnormal developmental syndromes. *Down's syndrome* (or Mongolism) for instance, with its characteristic facies and other features is now known to be due to the presence of three of chromosome 21 (trisomy 21) in every cell. Exactly how the extra chromosome or the sum of these chromosomes can produce the anomalies seen in the mongol child is unknown but the defect is probably an inhibition of development in specific regions such as the face. When the extra or missing chromosome is one of the sex chromosomes (X or Y), the defects, as might be expected, are usually in the development of the genitalia but not invariably. *Klinefelter's syndrome* (XXY) is an example of a trisomy of the sex chromosomes while *Turner's syndrome* (XO) is a condition in which one of them is missing. Another trisomy, XYY, is characterized by a tendency to tallness, aggressive and antisocial behaviour and mental defect but whether these features are always associated with the extra Y chromosome is still under investigation. Trisomies originate during meiosis in the development of the germ cell by non-dysjunction of the pair of homologous chromosomes, the result being that two instead of one chromosome pass into an otherwise haploid sex cell and at fertilization the normal accession of another chromosome brings the total to three (Fig. 15.1). Turner's syndrome (XO) and other deletions of a chromosome occur when, after non-dysjunction, there is a viable sex cell which does not receive its normal chromosome complement (one of each).

Anomalies in the number of chromosomes can be easily identified by examining them in special 'squash' preparations of cells in mitosis. Even missing or extra fragments in a chromosome, or transposition of a fragment from one chromosome to another can be identified microscopically but missing, extra or abnormal genes contained in the chromosomes cannot so far be recognized by this method. Yet these defects are probably far more numerous. To be sure that a congenital abnormality is the result of an inherited abnormal or defective gene or genes will depend on a careful history of the rest of the family i.e. parents, grandparents etc. as well as siblings (brothers and sisters) to find whether the same defect or a modification of it is present in a related member. An isolated defect of a gene, however, occurring in a newborn presents a very difficult problem in regard to its aetiology. Not all gene defects declare themselves in a family tree because abnormal genes (*abnormal genotype*) does not always give an abnormal end result to development (*phenotype*). It is known that normal and abnormal genes may be influenced in their activity by association or reaction with other genes (and not necessarily by the homologous genes) derived from the other parent. The normality or otherwise of development stemming from a defective gene is also related to whether or not it is sex-linked, i.e. the gene is in one of the sex chromosomes, (as in the case of haemophilia where only males exhibit the disease and females act as carriers) or whether it is dominant or recessive, homozygous or heterozygous.

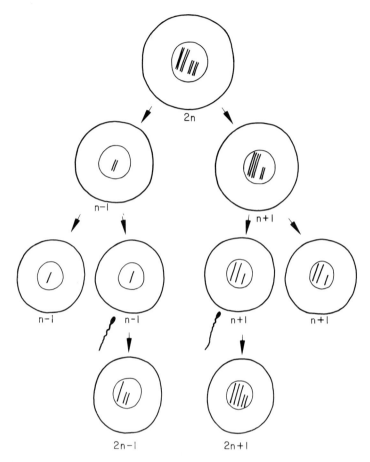

Fig. 15.1 Diagrams to illustrate how the simple trisomic condition and the monosomic condition arise during meiotic (reduction) division and fertilization. Only four chromosomes (2 pairs), each in replicated form, are illustrated in the parent cell.

ENVIRONMENTAL CAUSES OF CONGENITAL ABNORMALITIES

Equally important for the activity of such genes is the environment, especially nutritional, in which development proceeds and this feature does even more to confuse an already difficult problem. Environmental factors causing or contributing to the development of congenital abnormalities could all be grouped under the heading of *teratogens* although the term all too often suggests drugs only. The list of *possible* teratogenic agents is inexhaustible but the number of *known* teratogenic agents is appreciably smaller while *known* teratogens in the *human* number even less. In the first place, the site of action of teratogenic agents could be almost anywhere along the pathway of protein synthesis from the gene through transcription, translation and other steps to the final product interrupting or altering the process to bring about the abnormal result; or it may be associated with mitotic division which may also be inhibited and thereby arrest cell proliferation.

Ionizing radiations (e.g. X-rays), are well known for their teratogenic action. Because they often have their effect on the developing ovum or spermatozoon before it leaves the parent, causing damage or changes (mutations) in the gene, they are behaving as mutagens. The effect of such damage does not occur until embryonic development is in progress and the results may not come to light until birth but, more important, the effect may be transmitted with the mutated gene to succeeding generations. Radiation can also damage directly the genetic material in the proliferating cells of the embryo and fetus although in this case all cells in the fetus may not be affected and the mutation or genetic change will not be transmitted to the next generation unless the primordial germ cells or their descendants have been affected. At the molecular level, the ionizing radiation produces its mutagenic effect by upsetting the chemical structure of the DNA molecule, altering or inhibiting its activity in replication or transcription.

Other physical agents such as *heat, cold* or *trauma* have not been proven in their teratogenic effect on the human or other mammals but with ova and embryos which are not so sheltered, they have been shown to have definite effects.

Nutritional deficiencies (e.g. of vitamins), *toxins* and *drugs* are well known and feared as teratogens and we need only to be reminded of the effects of prescribing *thalidomide* during pregnancy to realize their dangers. Even the buffering action of the placental barrier and of the whole maternal metabolism does not safeguard the mammalian fetus and indeed the presence of these makes the detection of a teratogenic agent difficult. Thus, different animals have different detoxicating mechanisms or use different metabolic pathways and a substance ingested by one animal may damage the fetus because it has not been metabolized or broken down sufficiently or in a way which will render it innocuous. Conversely, many animals may metabolize a drug so efficiently that it has no teratogenic effect; if, then, these animals were the experimental targets or models in testing a particular drug and were found to give birth to normal offspring, no assumption can be made that the drug is safe for every animal. Such specificity of action was only too obvious with regard to thalidomide which had been tested in laboratories using the usual animals and shown there to be non-teratogenic. It was a shock to discover that it was having such a devastating but consistent effect on the human fetus and indeed comparable experimental results have only been achieved in some of the primates. All that has been said about the differing metabolic pathways in the maternal organisms can be repeated for the different placental barriers not merely because of the morphological variations but also in regard to their different physiological activities which control the passage of some materials to the placenta and also metabolise others on their way. Without a detailed knowledge of how the body metabolism and the placenta of every animal including man can deal with drugs, they must remain a danger to the fetus.

On the same theme, namely the difficulty of predicting the effect on the fetus of substances or drugs ingested by the mother, there is the major problem of identifying the biochemical reactions occurring at every step in the differentiation of every tissue in the fetus of each mammal. The importance of such information is exemplified by the extraordinary specificity of thalidomide; it attacks the development of the limbs in the human and near humans but its possible effects on other human tissues and on the limbs of other animals are not nearly so well defined. This underlines how little we know about the processes of tissue and organ differentiation, and the effects of teratogens may stimulate as well as be used in the further investigation of cell differentiation.

Another well-recognized cause of congenital abnormalities is *virus infection*, e.g. *rubella* (German measles) which produces a characteristic syndrome of anomalies in the heart, eyes and ears of the human fetus when the mother contracts this illness during the first three months of pregnancy. Probably the abnormality in development is caused by invasion of the fetal tissues and cells by the virus itself. As regards specificity, the same points should be emphasized in regard to teratogenic virus infections as with other teratogenic agents.

It is true that nutritional deficiencies, including the absence of certain substances e.g. vitamins and trace elements, from the diet and the presence of frank toxins which interfere with fundamental metabolic pathways in a cell may cause death of the fetus when the deficiency is severe or the dose high. This led to the view that any teratogen if administered in sufficient concentration would cause fetal death but the above discussion on the specificity of teratogens belies the truth of this sweeping statement. Many teratogens but not all, do come into this category. Even those which interfere with basic metabolic processes have a certain specificity depending on the exact time of administration, i.e. different tissues or organs will be affected according to the stage of development of the embryo. This is more readily understood when we remember that the differentiation of the various tissues and the emergence of the different organs in the embryo occur at definite times and in a fairly rigid sequence during development. Intense differentiation activity in a tissue or organ demands increased or accelerated metabolism in its cells and, if the essential metabolic pathways cannot rise to such needs because of some deficiency or toxin inhibiting them, then that particular developmental process may fail or lag. Other developmental processes which are, at that time, in a relatively quiescent phase of activity will survive since their metabolic needs are so much less. It is also true that failure of a developmental process because of an inhibitory or teratogenic factor cannot be compensated for at an appreciably later stage when normal conditions return. There is a time as well as a place for each developmental process in the programme. From all these teratogenic studies, it is clear that the effects can only be produced in the early stages of development i.e. when the tissues and organs are differentiating—corresponding to the first three months of human gestation.

In spite of all that has been learned from experimental work and from theoretical considerations of the etiological factors in teratology, the doctor still has an extremely difficult task to track down the cause of a congenital abnormality in a particular child unless it is easily identified as belonging to the group of well-documented syndromes, e.g. Down's Syndrome. It is the exception rather than the rule to find an excellent family history of the same disorder or to have reliable evidence of a teratogenic agent such as a drug, infection or nutritional disturbance being implicated. More often than not, intensive investigation of the isolated case provides no definite result. The identification of mass teratogenic activity, e.g. from thalidomide or rubella, unfortunately accounts for relatively few of the congenital abnormalities which the clinician meets and his main task remains the treatment of the affected child.

METABOLIC DEFECTS

There is still a tendency to think of congenital abnormalities in terms of structural deviations from the normal, and defects such as cleft (hare) lip, cleft palate, syndactyly, abnormal development of the heart, abnormalities of the kidney and urogenital tract, atresia of oesophagus, imperforate anus etc., are only too obvious; but there is an important group of defects which are fundamentally of a functional nature with perhaps nothing grossly anatomical or even histological to identify them. These are often referred to as *inborn errors of metabolism*; they arise in the same way as the structural abnormalities although more frequently from gene defects; these lead to the faulty production or absence of the normal proteins, secretions or enzymes; secondarily, other developmental processes are disturbed perhaps with structural changes. Phenyl-ketonuria, for instance, is a rare but well-known inborn error of metabolism in which there is virtually an absence of the enzyme phenylalanine hydroxylase for the digestion of the amino acid phenylalanine to tyrosine. The absence of the normal metabolites and the accumulation of phenylalanine and abnormal metabolites in the blood have a severe inhibitory effect on the development of the brain in particular. It is significant that early diagnosis of the condition followed immediately by rigid restriction of phenylalanine in the diet allows normal brain development to occur. Some relaxation of the strict dietary regime is possible after seven or eight years without causing damage to the brain. A number of other similar congenital errors in metabolism have now been recognized as the cause of severe mental retardation in children.

FURTHER READING

Allan J.D. & Holt K.S. (1965) *Biochemical Approaches to Mental Handicap In Childhood.* Edinburgh & London: E. & S. Livingstone Ltd.

Drillien C.M., Ingram T.T.S. & Wilkinson E.M. (1966) *The Causes and Natural History of Cleft Lip and Palate.* Edinburgh & London: E. & S. Livingstone Ltd.

Federman D.D. (1967) *Abnormal Sexual Development.* Philadelphia & London: W.B. Saunders Company.

Fraser F.C. & McKusick V.A. (1970) *Congenital Malformations.* Amsterdam: Excerpta Medica.

Fuhrmann W. & Vogel F. (1969) *Genetic Counselling.* London: Longmans, Green & Co. Ltd.

Holt K.S. & Coffey V.P. (1968) *Some Recent Advances in Inborn Errors of Metabolism.* Edinburgh & London: E. & S. Livingstone Ltd.

Langman J. (1969) *Medical Embryology* 2nd ed. Edinburgh & London: E. & S. Livingstone Ltd.

Moore K.L. (1974) *Before We Are Born.* Philadelphia & London: W.B. Saunders Company.

Robson J.M., Sullivan F.M. & Smith R.L. (1965) *Embryopathic Activity of Drugs.* London: J. & A. Churchill Ltd.

Rubin A. (1967) *Handbook of Congenital Malformations.* Philadelphia & London: W.B. Saunders Company.

Wilson J.G. (1973) *Environment and Birth Defects.* New York & London: Academic Press.

16 EMBRYOLOGY AND EVOLUTION

There is a close and logical connection between these two kinds of development although one is rapid and affects only the individual while the other is slow, occurs over many generations and influences whole populations.

The simplest way to understand the association is to see the evolutionary changes as the results of mutations which are transmitted to succeeding generations; in fact, these heritable changes are no more than deviations from the established pattern of development, i.e. congenital abnormalities. In this context, however, we should overlook these deviations from normal development which interfere with or are harmful to the survival of the individual; rather, we should identify and concentrate on features which are characteristic of evolution and beneficial to the individuals in a population. These should then be regarded as genetic mutations which, in the first instance, influence the course of embryonic and fetal development and are then inherited by the offspring.

Certainly there are areas of the evolutionary process where it is more difficult to apply this hypothesis and we can return to them after establishing the hypothesis with simple examples. For instance, the tail of the mammal disappears late in primate evolution with no sign of it externally in the adult human. Yet the early human embryo has quite a presentable tail comparing well with the corresponding stage in development of the lower animal. The difference lies in the amount or extent of its subsequent development, particularly of vertebrae and muscle; in the human there is little or no progress after the initial 'attempt' at growth which persists as the coccyx. In lay terms, this evolutionary step is considered as loss of the tail or its failure to develop, whereas in genetic/developmental terms, we see it as an irreversible and heritable inhibition of the activity of these genes responsible for the later development of the tail. Compare this with the failure or inhibition of growth in the facial maxillary process preventing it from reaching the median nasal process and the opposite maxillary process, giving rise to a cleft (hare) lip; this is a recognized congenital abnormality of development, in which there is (at least in many cases) a heritable inhibition or defect of the genes responsible for normal growth.

Inhibition of gene activity leading to inhibition of a particular growth process is the simplest illustration of the connection between development in the individual and evolution in a population. We can also appreciate from this example that each evolutionary step is, in the first instance, an alteration in the development of one particular part of the body or (and equally important) in the development of a particular function (cf. the inborn errors of metabolism).

Increase in or facilitation of the activity of a gene or group of genes producing thereby an increase in the size of a particular region or organ may be more difficult to understand as an evolutionary change. To illustrate the point, we should first consider the development of the various regions of the human face; the differences in individual faces depend on the relative amounts of growth in the different parts e.g. nose, cheek, lower jaw etc. Whether you have a large or small nose, for instance, will depend on the activity of the relevant genes which you inherited. It is a very exaggerated example to set against these minor variations but the trunk of the elephant is the result of intense activity by the genes responsible for the development of the nose and upper lip inducing more rapid and more prolonged growth of that region compared with other mammals; and the activity of these genes is out of line with that of the other genes controlling the growth of the face.

It is possible therefore to accept the exaggeration of a region or organ as increased growth controlled by mutated genes but there are still other variations in gene activity providing evolutionary changes, e.g. alterations in the timing of gene activity. Thus, in one species, a feature or organ may develop at quite a different time or rate in the programme of development compared with other species—so-called *heterochrony*. The best example of this phenomenon occurs in the development of reproductive ability compared with somatic development. In most animals, somatic growth and development are well advanced, perhaps almost complete before reproduction is possible; in amphibia, generally, the reproductive organs do not mature until after metamorphosis but in some newts these organs develop fully before metamorphosis (i.e. the gills are still present) and in the axolotl metamorphosis does not occur at all. This latter condition is referred to as *neoteny*, a special example of heterochrony. Of course, the argument may be advanced that it is somatic growth which is inhibited rather than reproductive maturation which is accelerated.

There are many other examples of neoteny such as the 'flat' face in man compared with the extended growth of the muzzle in other mammals and primates. Also, the foot of man is unlike that of other primates in having retained the embryonic condition with the big toe in line with the other toes, not opposable as it is in the foot of other primates and not comparable with the thumb of primates.

The kind of evolutionary change which has given rise to most discussion and is not so simply explained, is exemplified by the different end results of development in the visceral arch region. In fish, for example, the arches and grooves (or pouches) become gills and gill clefts whereas in mammals, development in this region is diverted or deviated into other pathways leading to the emergence of entirely different structures. This kind of evolutionary change stimulated the belief that ontogeny (development of the individual) recapitulates phylogeny (evolution). Mistakenly, however, it was assumed that the development of the neck in mammals began with a pattern representing the entire development of that region in a fish and was followed by the pattern of mammalian development. Much nearer the truth is that evolutionary changes originating at the gene level were responsible for altering development *after* the formation of the arches and grooves. In the mammals, completion of the fish pattern of development was not achieved. Exactly what happens at the gene level is difficult to establish; possibly there have been many mutations in the course of evolution affecting visceral arch development but only in its later stages and none of them influencing the early basic steps.

A great deal more time could be spent discussing evolutionary changes such as the appearance of an apparently new feature in the life cycle, e.g. the placenta. But that organ is not such a new and unexpected development as it appears. For instance, the chorioallantois of the cleidoic egg can easily be visualized as the precursor of the chorioallantoic placenta while the yolk sac foreshadows the yolk sac placenta. And looking even further back in phylogeny, we also find a close association between the blood vessels of the yolk sac and the maternal tissues, e.g. uterus, ovary, in those amphibia and fishes which are ovoviviparous. The emergence of the mammalian placenta, therefore, could quite easily be regarded as the culmination of a series of advantageous mutations.

We could summarize the views on evolution as expressed here by saying that mutations which alter structure or function in the body of an animal advantageously will not only persist but flourish in a population. A mutation is, of course, an advantage only if it helps the animal to derive more benefit or survive more easily in its environment. But mutations may be quite widespread in a population and of no advantage until a change in the environment occurs e.g. alterations in temperature, availability of food or type of food, appearance of new predators and new infections etc., etc. Only then may the mutation become advantageous by helping the animal to adapt to and survive in the different environment.

FURTHER READING

De Beer G. (1958) *Embryos and Ancestors* 3rd ed. Oxford: Clarendon Press.
Dobzhansky D. (1967) *Evolution, Genetics and Man.* New York: John Wiley & Sons, Inc.
Hamilton T.H. (1967) *Process and Pattern in Evolution.* London: Collier-Macmillan Limited.
Savage J.M. (1963) *Evolution.* New York: Holt, Rinehart & Winston.

17 BEHAVIOUR AND LEARNING

The development of behaviour and the ability to learn, leading in the end to the mature personality, is not always considered as an integral part of general growth and development. Too often, it is regarded as a separate topic, unrelated even to the development of the nervous system and associated more intimately with education and the environment. Behaviour and learning must have a structural (gross and microscopic) basis as well as a physico-chemical one and should be related to these, but our scanty knowledge of the development of the nervous system at the cellular and molecular level makes it difficult to do this.

DEVELOPMENT OF THE NERVOUS SYSTEM

The following concept of the nervous system and its function can be criticized on the grounds that it is too simple but at least it is a basis for a hypothesis about the development of behaviour and learning. The functional unit is the *neurone*, comprising a cell body, its axon and its dendrites, and these units 'communicate' at *synapses* where the axon of a neurone touches the dendrite or cell body of an other. Groups of sensory neurones are responsible for collecting information obtained by stimulating of their dendrite endings, e.g. by touch, light etc. from the environment, passing these nervous impulses, directly or indirectly, to motor neurones which, via their axons effect a reaction by the animal to the initial environmental stimulus: the simplest reaction (reflex) is the contraction of a muscle or muscles to move the limb or body away from some noxious stimulus. The interpolation of the internuncial neurones between the sensory and motor pathways provides a vast number of possible variations of the simple reflex; thus, impulses may be inhibited, strengthened, distributed to other circuits or channelled along particular pathways. The establishment of *integrating centres*, i.e. aggregations of internuncial neurones, where impulses (information) are collected from a variety of different sensory pathways, correlated, sifted and redistributed to an equally great variety of different motor pathways, gives discrimination in reactions to the environment and leads to the development of a brain as we understand the term. These are the gross morphological features of the concept but other details of the scheme are necessary e.g. the development or establishment of the connections (synapses) between the neurones, and there is more to their development than merely the approximation of parts from different cells. There is no continuity of cytoplasm or even contiguity of the two neurones because examination of the synapse by the EM shows a minute but definite gap between the two cells. The function of a synapse i.e. the transmission of an impulse, is associated with a chemical substance (transmitter), e.g. acetylcholine and the presence of very small cytoplasmic vesicles in the axonal component of the synapse; these vesicles probably do not appear *immediately* after the synapse has been established. It is quite possible that function (dependent on the presence of the vesicles) is induced by stimulation of the neurone on the afferent side, i.e. an initial environmental stimulus or perhaps even repeated stimuli are necessary. Full development of all the dendritic processes and therefore the development of all the interneuronal connections is known to continue long after the basic plan of the nervous system is established. Obviously, then, all the delicate variations in how the impulses coming into an integrating centre may be dealt with must await the complete morphological development of all connections. Following on that, there is of course the functional development and, if stimulation plays a part, further consideration and investigation in that direction is needed. Channelling impulses and information along the correct pathways into, through and out of integrating centres giving, tor instance, the fine discrimination of muscle movement in a limb are details of which we have very little knowledge so far. The synapses may be the main agents but the myelin sheaths of nerve fibres may also play a significant part—if nothing more than as insulators capable of isolating impulses to the individual fibres—but that may be too simple a concept. However, myelination of nerve fibres is not complete until an appreciable time after birth in most animals including the human and full development of function in the nervous system can, in some measure, be related to myelination.

Function and activity in the fetus and child

With this picture of progressive development in the nervous system before us, it should be possible to acquire at least some understanding of how a baby develops in regard to its reactions and behaviour. The first signs of activity in the nervous system appear very early in fetal life and the increase in number and complexity of these signs or responses to environmental stimuli during development form a pattern which is, within limits, consistent. Remember that, *in utero*, the fetus in spite of being bathed in amniotic fluid is not by any means isolated from environmental stimuli; its own limbs, the wall of the uterus, the umbilical cord and even sound, according to some authorities, can stimulate its sensory neurones.

Tests on the live human fetus removed surgically from the uterus for therapeutic reasons have shown that, as early as $8\frac{1}{2}$ weeks of age, it can open its mouth in response to touching that area of its face and at a little later stage in development the limbs can be made to respond to similar stimuli; as the fetus grows, more and more vigorous responses can be obtained, each reaction involving widespread movement, e.g. the whole of one side of the body. There is, particularly with the earliest responses however, evidence of a progressive development of function in a cephalocaudal direction and presumably therefore of the associated structural developments as well in the nervous system. *Quickening*, the movements of the child *in utero*, felt first by the mother about mid-pregnancy are presumably reactions on the part of the fetus to mechanical stimulation from its own environment. But the best known signs of functional development in the nervous system are seen from the 28th week of fetal life onwards—the earliest stage at which a newborn child is legally viable, i.e. thought capable of surviving. (Babies have survived from earlier stages however when careful and sustained treatment has been provided.) At 28 weeks many reflexes can be elicited although there is still little sign of localization in the response, e.g. a 'grasp' reflex by the hand can be obtained by gentle pressure on the palm, and the arms and legs respond wildly to various stimuli. On the whole however the child is limp and 'floppy' with no 'tone' or tightness in its muscles, i.e. the arms and legs can be passively moved through a much greater, even frightening, range than is possible at birth; nor is the child alert to its surroundings except for very brief periods but as it progresses to 40 weeks (full time) it becomes far more awake and reactive to its environment for longer spells. Its arms and legs, instead of lying limp at its sides or wherever they happen to fall, become flexed, drawn up across the chest or in front of the abdomen. At this stage it is difficult to stretch them and obviously nerve impulses must be causing the flexor muscles to contract very strongly. Those responses which could be obtained weakly at 28 weeks are brisk and more defined; for instance, the 'mouthing' reflex is strong and prompt whilst the response to pressure on a limb, apart from being brisker is likely to be confined to that limb instead of uncoordinated movements of all of them. A response to light can be identified and the eyes move together but recognition of detail by sight comes later when the visual pathways are fully matured.

After birth, the extensor muscles of the neck, trunk and limbs, gradually and in that order, acquire the degree of tone which the flexor muscles had at birth. Only then can the child hold up its head, straighten its back and eventually sit up (Fig. 17.1). It should be obvious that there is progressive development in the neuronal connections both structurally and functionally during those first six months of post-natal life. The same is true and better exemplified when it comes to crawling and walking; by a year old, with the child's first unsteady steps on its own, not only are the muscles strong enough but their activities are now being co-ordinated by the nervous system to accomplish the feat of walking. And no amount of dedicated teaching or training will make a child walk earlier than its own inborn schedule allows because the nervous system is not sufficiently developed in detail to achieve these activities. Yet we know that after the sixth month of intra-uterine life there is no further increase in the number of nerve cells. Further development from then onwards consists of increasing the number of dendritic processes from the neurones, establishing fresh synapses, initiating function in them and myelinating the nerve fibres. A further illustration of progressive maturation in the nervous system can be seen in the way a child uses his hands; picking up an object is at first a clumsy grasp but this finally gives way to dealing with it in a more delicate fashion with forefinger and thumb. Also, there is the ability to speak which involves refinement in certain sensory, e.g. hearing, pathways but the motor co-ordination of lips, tongue and larynx is an other example of the development of organized activity in several parts of the nervous system. The whole picture of behaviour and activity in a child as it grows has a background of progressive involvement of the muscles in a crude fashion overlain by progressive refinement of movement and manipulation. Important, too, are the pattern and progress of the development of the child in these respects, e.g. crawling, walking, speaking. The order in which the responses and faculties develop is reasonably but not absolutely consistent; the rate at which they appear varies enormously even in normal children.

Learning and intellect

If the child's progress throughout the maturation of motor activity is poorly understood, then the development of its learning and of its more intellectual activities is even more obscure; this field of development has few tangible signs or definite morphological features but it is now recognized that the acquisition of certain intellectual facilities are characterized by and related to 'critical periods' during which the child learns the skill or masters the intellectual process in question. Before the critical period, little or nothing can be achieved by coaching or teaching the child that skill or intellectual process. The period from 2 to 6 years of age is that in which language and verbal communication are most easily learned; on either side of that period, before and after, there is a marked decline in aptitude towards learning that facility, and it is well known that a child can very quickly learn a new language whereas the adult may find it very difficult. It is very tempting to accept the view that, during the critical period the different nervous pathways and integrating centres required for linguistics are maturing or have just matured in structural terms; they are then at their most receptive to stimulation and the stimulus of hearing the spoken word, repeated several times, induces full function in the relevant synapses. From a simple though perhaps naive concept such as this we can visualize a succession of intellectual facilities being acquired, as long as the relevant environmental stimuli synchronize with structural maturation. In the development of the intellect there is also the need for memory and reasoning. Memory is the ability to recall what has been experienced or learned at some earlier date. The stimulus which evokes recall must, within the nervous system, be able to associate with the pathway(s) initially involved in learning or experiencing the original facts or events. Facilitation or the ease of transmission of impulses along these pathways will depend on many factors, e.g. degree of repetition of the initial stimulus, the concentration or attention directed towards the fact(s) or incident at the time and the age of individual at that time. Many views have been

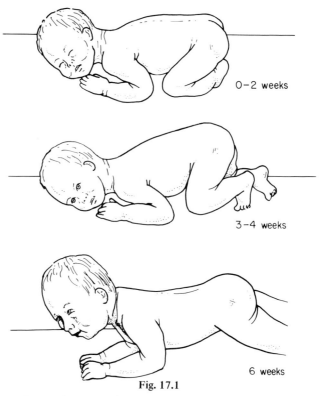

Fig. 17.1

expressed on the mechanisms whereby recall can occur but, until we can better appreciate the structural and functional organization in synapses, and particularly in those within the integrating centres, memory will be a challenging problem. In the context of the developing nervous system, it is of interest that incidents experienced and facts learned in childhood and adolescence are usually more easily remembered by the adult than are more recent episodes and information.

Simple reasoning and deduction are faculties acquired by the child between the ages of 7 and 11 but the extension of these to the solution of problems in the abstract comes only in the next few years. Perhaps there are no new structural or even physicochemical changes in the brain so late in life to account for these but again there is a temptation to postulate the formation of more synaptic connections or, at least, that they only become functional at that time with experience and training. A great deal of observation and investigation has been devoted to the intellectual development of the child with little reference to and less knowledge of the actual stages in the development of nervous tissue but an understanding of the child's progress in this sphere comes more easily if there is even a speculative plan of structural/functional development at the cell level.

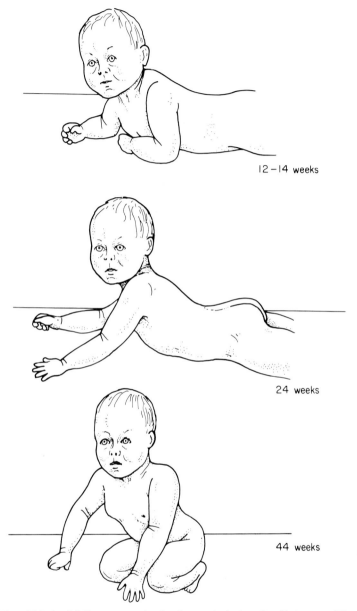

12 – 14 weeks

24 weeks

44 weeks

Fig. 17.1 A child's progress in development during the first year. After Illingworth, R.S., *The Development of the Infant and Young Child*, 3rd edition. R. & S. Livingstone Ltd., Edinburgh.

Behaviour and personality

Concerning the less tangible aspects of psychological development, viz. temperament and personality, there is even less opportunity to relate the child's reactions and behaviour to the development of the brain. The emergence of fears, moral and other standards, anxieties, motives etc. has nevertheless been shown to be associated with the age of the child i.e. critical periods, some features appearing in the first eighteen months of life others between eighteen months and three years and so on during the first decade. In some instances, it is just possible to relate details tentatively to the development of the nervous system. Much more attention has been given to the environmental effects such as teaching, training and experience as 'inducing' agents or stimuli for the development of these characteristics of a child's behaviour and no one will deny that the environment has an important role to play.

How is it that children and adults vary so much in learning skills and behaviour even when they have experienced identical teaching, training and environment? Naturally we can attribute this to the same genetic variation which all of us show, e.g. the hands and faces of members in the same family are never identical (except in identical twins and sometimes not even in them). An even better illustration is provided by our fingerprints. We should realize then that no two brains are likely to be identical, particularly in the details of interneuronal connections and even in the number of nerve cells. The functional capacity of each brain is therefore bound to be different from every other and, as long as we can accept that fact, we can appreciate that there must be a very wide spectrum of intellectual ability, primarily dependent on inherited structural features and capable of being modified only within limits. It may be a ridiculous comparison to make but the champanzee and bigger apes, although their brains are similar in shape and gross organization to the human brain, cannot be trained or taught to anything approaching the standards of the dullest child.

FURTHER READING

Allan J.D. & Holt K.S. (1965) *Biochemical Approaches to Mental Handicap in Childhood*. Edinburgh & London: E. & S. Livingstone.

Barnett S.A. (1962) *Lessons from Animal Behaviour for the Clinician*. London: William Heinemann Medical Books Ltd.

Barnett S.A. (1973) *Ethology and Development*. London: William Heinemann Medical Books Ltd.

Falkner F. (1966) *Human Development*. Philadelphia & London: W.B. Saunders Company.

Greenough W.T. (1973) *The Nature and Nurture of Behaviour—Developmental Psychobiology*. Readings from *Scientific American*. San Francisco: W.H. Freeman and Company.

Illingworth R.S. (1970) *The Development of the Infant and Young Child* 4th ed. Edinburgh: E. & S. Livingstone.

Manning A. (1972) *An Introduction to Animal Behaviour*. London: Edward Arnold.

Mark R. (1974) *Memory and Nerve Cell Connections*. Oxford: Clarendon Press.

Moltz H. (1971) *The Ontogeny of Vertebrate Behaviour*. New York & London: Academic Press.

Tanner J.M. (1961) *Education and Physical Growth.* London: University of
    London Press Ltd.
Weihs T.J. (1971) *Children in Need of Special Care.* London: Souvenir Press.

18 DEVELOPMENT IN PLANTS

With plants and animals looking so different and with so many obvious differences in the way they develop, it is surprising how closely the basic processes of growth and development in these two great divisions of the living world resemble each other. Comparisons and even contrasts between animal and plant development deserve attention because they provide a better understanding of development generally and may even help solve some of the problems in this field.

As in the case of animals, it is impossible to treat the development of every plant and still hope to present a sensible pattern of development. For many reasons, the flowering plant is the best example to describe and to offer as a pattern although there are numerous features which are peculiar to that kind of plant. The most convenient point to 'break into' its life cycle is the stage of the flower, the part of the plant which is specialized for reproduction. Fig. 18.1 knows the main parts of a flower; the *receptacle* or *torus* carries the whole flower including *sepals* and *petals* and particularly the pistil; this is the flask-shaped structure in the centre, consisting of an expanded basal part, the *ovary*, and a long hollow slender *style* with the stigma at its tip; around the style are

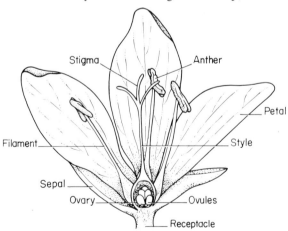

Fig. 18.1 The parts of a flower. From Greulach, V.A. and Adams, J.E., *Plants. An Introduction to Modern Botany*, 2nd edition. John Wiley & Sons, Inc., New York.

the *stamens* each comprising an *anther* carried on the tip of a thin *filament* arising beside the ovary.

The ovary contains one or more compartments and, in each, one or more *ovules* attached by a stalk to the placental area of the wall of the ovary (Fig. 18.2). The mass of cells comprising the ovule is the *nucellus* (or megasporangium) which is almost completely enveloped by tissue, comprising the one or two *integuments*, growing up from the base of the nucellus and leaving only a tiny *micropyle* (opening) leading into the nucellus (Fig. 18.3). One cell, the *megaspore parent cell*, in the nucellus undergoes meiosis to produce four cells, only one of which, the *megaspore* survives and increases rapidly in size. Its nucleus divides to give eight nuclei, some of which form separate cells, and one

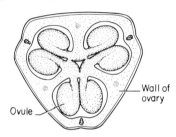

Ovule

Wall of ovary

Fig. 18.2 An example of the arrangement of ovules within the ovary. From Greulach, V.A. and Adams, J.E., *Plants. An Introduction to Modern Botany*, 2nd edition. John Wiley & Sons, Inc., New York.

of these becomes the *ovum* or *megagamete*; two of the nuclei, however, (the *polar nuclei*) remain free but all are enclosed in the original cell wall of the megaspore (the *embryo sac*). Meanwhile (Fig. 18.3), in the anther (microsporangium) the *microspore parent cells* divide meiotically to give *microspores*; each microspore divides to form two cells of unequal size adhering to one another, the small *generative cell* and the large *tube cell*, together known as a *microgametophyte* or *pollen grain*. The pollen sac containing the pollen finally ruptures to release the pollen grains and these find their way to the stigma of the same or another flower (*pollination*). There the part of the pollen grain containing the tube nucleus i.e. the tube cell, elongates and burrows down through the lumen of the style into the cavity around the ovule, through the micropyle to reach the embryo sac. The other nucleus divides to give two *microgametes* or *sperms* which pass along the tube cell and enter the embryo sac. One sperm fuses with the ovum (*fertilization*) to form the zygote while the other fuses with the two polar nuclei to provide the triploid endosperm nucleus. Subsequent repeated divisions of this nucleus usually, but not invariably, with the development of a cell wall around each of the nuclei is followed by their aggregation around the zygote; these form the *endosperm*, a large mass of cells or cellular material which accumulate food from the parent plant to provide for growth of the embryo. A *seed* has now been formed, comprising zygote, endosperm and the enveloping integuments which become modified, according to the plant, to become the *seed coat*; it may, for instance

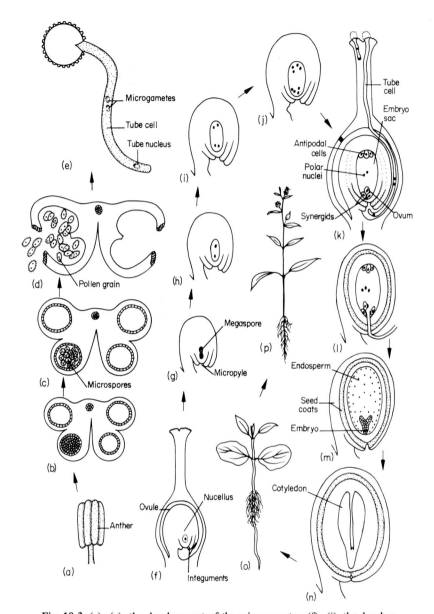

Fig. 18.3 (a)–(e), the development of the microgametes; (f)–(j), the development of the ovum; (k)–(p), fertilization, embryo development, the seed and early development of the plant. From Greulach, V.A. and Adams, J.E., *Plants. An Introduction to Modern Botany*, 2nd edition. John Wiley & Sons, Inc., New York.

be paper thin, fleshy or tough and cutinized. We can refer briefly, at this stage to the fortunes of the seed as a whole. The wall of the ovary may simply dry up and discharge its seeds to scatter or be distributed by some other means, finally settling in soil to germinate when environmental conditions are favourable. As long as the seeds are within the ovary, these, i.e. seeds and ovary, constitute a 'fruit' although this is not the common interpretation of the term. The layman's idea of a fruit is exemplified (Fig. 18.4) by the tomato in which

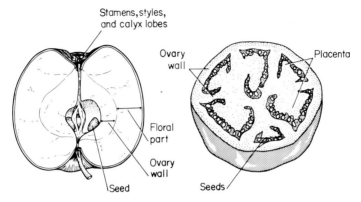

Fig. 18.4 An apple and a tomato cut to show the seeds and the origins of the different parts of the 'fruit'. From Greulach, V.A. and Adams, J.E., *Plants. An Introduction to Modern Botany*, 2nd edition. John Wiley & Sons, Inc., New York.

the seeds are embedded in the soft fleshy tissue developed from the ovary, or by the apple in which the seeds in their tough seed coats are buried in a mass of fleshy tissue formed by the wall of the ovary and around that by similar tissue derived from the receptacle and adjacent parts. In the strawberry, the fleshy tissue is provided chiefly by enlargement of the receptacle which carried the seeds on its surface and in nuts ('fruits' nevertheless) the ovary wall has become almost stony hard. These modifications of course help in the dispersion as well as the protection of the seeds.

Meanwhile the zygote undergoes cleavage and proceeds to the development of an embryo. Invariably, it proceeds only so far, ceases to develop for a varying period and then resumes; and each species has its own particular variation of this pattern. At its first cleavage, the zygote divides unequally into a large and a small cell, the latter being directed into the developing endosperm mass, the former lying near the edge of the embryo sac. Furthermore, the large cell divides repeatedly to form an elongated stalk or *suspensor*; the small cell by its subsequent cleavages is mainly responsible for producing the embryo itself, developing into a small mass, at first globular and later of a conventional heart shape with the narrow, more pointed end attached to the suspensor. There is little or no movement of cells as their numbers increase but it is possible to indicate where the individual parts of the later embryo and young plant will appear. The root will develop in the region adjacent to the suspensor; the cells in the slight depression at the other end will give rise to the stem while on either side of that region, the elevations are to form the cotyledons.

At or about this stage, the embryo stops developing and the whole seed enters a stage of dormancy (see later); only when this stage is broken and usually associated with germination does further activity resume. The radicle extends out from the narrower end; it has a cap of cells covering a group of actively dividing cells which produce the tissues of the root. (The emergence of the root or radicle through the seed coat is usually the first outward sign that germination is occurring.) The actively dividing cells near the tip of the root form the *root meristem* (Fig. 18.5) and retain this potential to proliferate

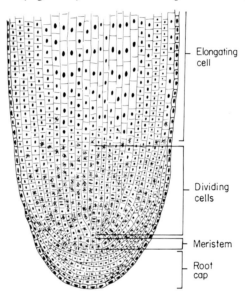

Elongating cell

Dividing cells

Meristem

Root cap

Fig. 18.5 Longitudinal section of root to show root cap, meristem and the differentiation of cells derived from the meristem. From Greulach, V.A. and Adams, J.E., *Plants. An Introduction to Modern Botany*, 2nd edition. John Wiley & Sons, Inc., New York.

throughout the life of the plant; the cells derived from them continue dividing for some time after leaving the meristem but soon differentiate into the specialized tissue cells of the root (Fig. 18.6). The most obvious feature in the cells as they differentiate is elongation, e.g. in the conversion to sieve-tube elements which, end-to-end, form food-carrying channels, the first signs of the phloem. In much the same way, other cells (also derived from the meristem) form the tracheids or vessel elements of the xylem. Also derived from the meristem is the *vascular cambium*, at first a relatively undifferentiated tissue lying alongside the xylem and phloem. By cell division and further differentiation this tissue provides secondary, i.e. additional, vascular channels (both xylem and phloem) and hence increases the thickness of the root.

At the other end of the embryo, the cotyledon(s) grow to form the first leaves of the new plant but they may play a significant part in food storage. In some seeds, the mass of endosperm surrounding the embryo remains in that position and is used by the embryo as required. In other cases, the cotyledon

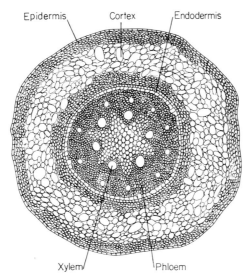

Fig. 18.6 A transverse section showing the differentiated tissues of the root. From Greulach, V.A. and Adams, J.E., *Plants. An Introduction to Modern Botany*, 2nd edition. John Wiley & Sons, Inc., New York.

becomes thick and fleshy in consistence by assimilating and storing the food material of the endosperm tissue, thus incorporating it for use at a later stage.

Between the cotyledons, there is a small group of cells, the *apical meristem*, which by proliferation produces cells to form the stem in the same fashion as does the root meristem. These newly formed cells of the stem soon stop multiplying, begin to elongate and differentiate into sieve tube elements for the phloem and tracheids for the xylem; a vascular cambium is also formed to add to those vascular channels during growth in thickness of the stem at later stages. As with the root in its development, other parenchymatous tissues in the stem also develop from the meristem. In this way, generally, the meristem rises on the stem which it has produced. The cells of the meristem retain their 'embryonic' behaviour and subsequent growth in the height of the plant depends on their continuing to proliferate (Fig. 18.7). Dotted around the sloping side of the meristem, always in a very definite pattern, there are little elevations caused by cells proliferating and later differentiating to become leaves; as the meristem continues to produce the stem (Fig. 18.7), the leaves drift apart from one another and find themselves growing out from the stem. In the acute angle between leaf and stem, some distance away from the tip of the meristem, a small group of meristematic cells can sometimes be identified: these give rise to *lateral* or *axillary* shoots and, for their subsequent growth, their meristems must retain potentially if not actively growing cells. Seasonal growth in plants which add to their height each year is dependent on their meristems resuming their activity in spring after a period of winter dormancy. The meristem is seldom exposed to the environment; even when in active growth it is well

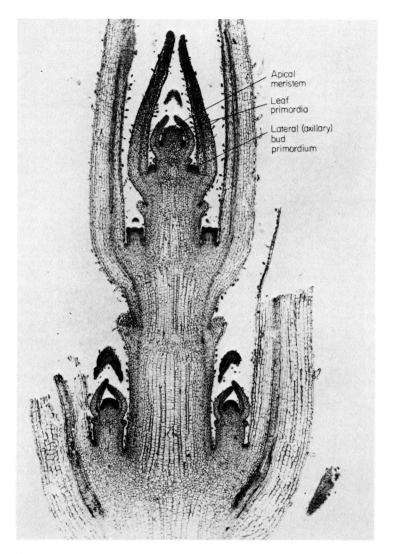

Apical
meristem

Leaf
primordia

Lateral (axillary)
bud
primordium

Fig. 18.7 Longitudinal section through a bud, showing meristem, leaf primordia and developing leaves. By permission of Dr. F. Cusick.

protected by the close approximation of the new developing leaves which form a bud around it (Fig. 18.7). Notice, too, that there is no sign of active migration of any cells or cell groups during the development of any of its parts; cells are pushed away from their site of origin, multiply and grow in length wherever they find themselves.

The meristem at the tip of a plant and those on the lateral shoots are capable of giving rise to flowers. The mechanisms by which the meristematic cells change from their normal activities to become flower-forming tissues are obscure.

Instead of proliferative growth giving rise to a stem and leaves, there is a change of growth and differentiation patterns to produce sepals, petals and the reproductive organs in that order from the edge of the meristem to its tip. The meristem cannot revert to its original function after it has embarked on flower development and when the flower dies the meristem is no longer there.

The development of plants looks deceptively easy until you begin to ask questions about the mechanism controlling the development of flowers or the factors involved in dormancy for instance. A great deal has been achieved in this field but the picture is still far from clear. Such problems cannot be discussed at length in this text and we can only hope to introduce them to allow contrast and comparison with animal development. Furthermore, the inter-relationships of the various factors involved and the overlapping roles they play during plant development and behaviour make it difficult to give a coherent story without repetition.

It is well known that plants are influenced by light; for instance, there are some which come into bloom in early summer when, in temperate climates, the daylight is increasing whereas others come into flower in the autumn as days are shortening. In other words there are *long-day* and *short-day* flowering plants but there are also others (*neutral-day*) which behave independently of the period of light. Investigations into this phenomenon have revealed that the length of the period of continuous darkness is more significant than the length of exposure to light and that interruption of the dark period may destroy the effect. Thus there is a critical length to the period of darkness for stimulation of flowering, acting primarily on a pigment, phytochrome, in the leaves. From these, a stimulus is transmitted to the meristems causing the changes leading to the development of flowers, and the existence of a hormone, *florigen*, has been postulated for this step; the substance has not been isolated or positively identified but the evidence for it is reliable. For instance, only one leaf need be kept for the necessary time in the dark to bring on flowering and even a leaf from a treated plant, grafted to an untreated stem, may induce flowering in the host. A period of light is of course essential for ordinary photosynthesis to nourish the plant in the normal way. Unfortunately, like so many other inductive reactions in biological growth, the precise inter- and intra-cellular processes in the meristem are still obscure. The phenomenon discussed so briefly here is known as *photoperiodism* but light is also responsible for *phototropism*, i.e. the tendency for a growing plant to bend towards the source of light; the mechanism is discussed below in relation to the hormone, auxin. A somewhat similar phenomenon is *geotropism*, in which the growing root of a plant always turns downwards while the stem turns in the opposite direction, both occurring irrespective of the initial orientation of the plant. *Regeneration* on which a great deal of work has also been done on animals is a prominent characteristic of plants. Dedifferentiation seems to occur more frequently and more readily in the plant cell than in the animal cell and this may be of help in elucidating the problems of dedifferentiation and regeneration in its wider sense.

Dormancy, mentioned earlier in relation to seeds and buds, is an expedient of prime importance for resisting harsh environmental conditions such as cold and drought. It refers to the cessation of growth in the embryo on the one hand and at the meristems on the other. The main environmental factors involved in dormancy are light and temperature but, in fact, the latter is more often the means whereby dormancy is broken (and germination initiated) rather than the cause of it. The essentials for breaking dormancy also include the presence of moisture which may play a part in softening the seed coat as well as helping the embryo to resume its growth; hormones, too, are important in the maintenance and in the breaking of dormancy.

Vernalization, a related phenomenon, is the application, natural or artificial, of cold ($0-5°$C.) to the seed or to the plant at a later stage in development before it can come into flower. Unlike the light/dark stimulation (photoperiodism) for flowering, the sensitive region of the adult plant is the meristem itself. Although grafting experiments in some species have transmitted the stimulus for flowering from vernalized to unvernalized plants and led to the view that a hormone 'vernalin' is involved, some species, e.g. chrysanthemum, require each growing tip to be vernalized and fail to transmit the stimulus in grafting experiments; this suggests that the stimulus is localized to the vernalized cells and their descendants and is in keeping with the behaviour of plants which can be vernalized at the seed stage.

Senescence is of widespread occurrence in plants, e.g. in leaves and even in whole plants such as annuals. A plant hormone, abscisic acid, has been identified with this process as well.

Apical dominance, one of the most interesting phenomena in plant growth, led to the identification of the best-known group of plant hormones viz. the *auxins*. It was known for a very long time that removal of the apical meristem in a plant allowed the highest lateral (axillary) shoots to grow unusually long, i.e. their meristems were stimulated to grow or some kind of natural inhibition on them was removed (Fig. 18.8). Replacement of the apical meristem with a small piece of agar containing auxin prevented the outgrowth of these lateral shoots. The main source of this hormone is the apical meristem and young leaves within the apical bud, although the individual cells responsible have not been identified. From its source, the auxin travels down through the stem as far as the root but the exact pathway it takes is not clear; nor has it been established how the auxins can under normal circumstances inhibit proliferation and outgrowth of the lateral shoot meristems, although these meristems themselves produce a great deal of auxin if they are liberated from the dominance of an apical shoot. A concentration gradient of the auxin occurs in plants—high at the apex, low in the root. There are several synthetic as well as natural auxins, all related to the best-known, i.e. indole-3-acetic acid, and apparently the action of these substances is stimulation of growth, i.e. enlargement of the cells by increasing the plasticity and allowing stretching of the cell walls.

Phototropism is believed to be mediated through unequal distribution of auxin across the stem; it is found to be in a higher concentration on the side of the plant opposite the source of light, causing more extension or growth of the

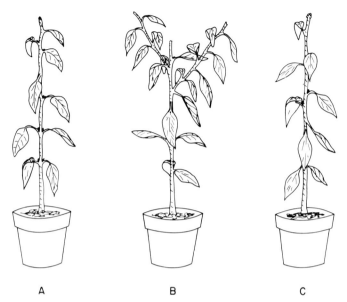

A   B   C

Fig. 18.8 (A) Normal plant showing terminal bud; no branching. (B) Removal of terminal bud abolishes apical dominance and allows growth of branches from lateral buds. (C) Substitution of terminal bud by paste containing auxin restores apical dominance and prevents growth of lateral buds. From Greulach, V.A. and Adams, J.E., *Plants. An Introduction to Modern Botany*, 2nd edition. John Wiley & Sons, Inc., New York.

cells on that side and bending of the stem towards the light. Whether this imbalance in the distribution of the auxin is caused by destruction of auxin by the light in the tissues nearest the light or by migration of the auxin away from the light has not been established. Geotropism is also largely attributed to the action of the auxins. It has been suggested that elongation of the cells on the lower side of the stem while it lies horizontally is due to the auxin collecting or seeping down by gravity to that side and stimulating growth in the cells there to turn the plant upwards. It would be difficult to use the same theory to explain the downward growth of roots when placed horizontally because the cells which are elongating are those on the upper side of the root. Nevertheless, the phenomenon is attributed to auxin activity on the grounds that the normal very low concentration required for root growth is, under these circumstances, unevenly distributed across the root and, when present in high concentration on its lower side, it is actually inhibiting growth and creating a differential in growth. How the auxins actually stimulate growth in cells has not been established but one view is that they increase the synthesis of the enzyme(s) which can digest some of the constituents in the cell wall thereby increasing its plasticity. Auxins have numerous practical applications; synthetic preparations are used to stimulate rooting, i.e. to increase the rate of root development in cuttings; in the opposite direction, auxin preparations, in excess of what is normally available to weeds, are applied as weed-killer sprays to distort and destroy normal growth in these plants. Since auxins are contained in pollen, the hormone is believed to be responsible for initiating the development ('setting') of fruit and indeed fruits have been 'set' in the absence of pollen (parthenocarpy) by the application of auxin. The same hormone coming from seeds is credited with maintaining the growth of fruit and here, too, auxins have found a practical application. There is considerable evidence however that auxins and another group of growth hormones, the gibberellins, collaborate to provide many of these effects particularly in relation to fruit development and growth.

The *gibberellins*, all chemically related, are produced by the cells of the apical meristems and of the leaves. They increase the internode stem length of plants, often dramatically and particularly in dwarf varieties. Their mode of action is not yet clear but the end result is both elongation and multiplication of cells in the stem. Although in many instances gibberellins and auxins are complementary in their activities, gibberellins are growth-promoting substances in their own right and have effects not possessed by auxins. For example, the breaking of dormancy in seeds and buds is attributed to the increasing accumulation (production) of gibberellins in these; and gibberellins, unlike auxins, inhibit rooting at all concentration levels.

The *cytokinins* stimulate cell division and differentiation. Again, there is evidence that these substances work closely with auxins, each potentiating the effects of the other. They are purines or purine derivatives artificially prepared but substances derived from plant tissues, e.g. from endospermic substances like coconut milk, have been shown to have the same effects.

The gas, *ethylene*, is now considered important in the ripening of fruit by stimulating respiratory processes but its activity is, like those of other hormones, closely associated with that of auxin which may be responsible for ethylene production in many different plant tissues.

There is only one well known growth inhibiting substance, namely *abscisic acid* and it is of fairly widespread occurrence. Apart from its effect in causing abscission of leaves, a phenomenon which can be inhibited by the application of auxins, abscisic acid, after being produced in the leaves, travels to buds on the plant, accumulates there and induces dormancy. Seed dormancy is sometimes characterized by a high concentration of abscisic acid in the seed coats and in the embryo. Breaking of dormancy is associated with an increased concentration of growth-promoting substances, e.g. auxins and gibberellins, but whether there is accompanying destruction or leeching out of the growth inhibitor to allow growth promoters to act is not clear. Abscisic acid is believed to produce its effects on cells by inhibiting the transcription of RNA.

The absence of cell migratory movements (a common feature in animal development), the lack of endocrine organs as such and its different type of vasculature have not prevented the plant from using hormones for affecting tissues at a distance from their source or for inducing and altering cell differentiation in special regions. Fundamentally, growth and development in animals and plants comprise the multiplication of cells, followed by their differentiation and the emergence of tissues and organs. The genetic influence is there in each case but its mode of expression is no clearer in one than in the other; no one can deny that primary inductive phenomena, so well established in animals is absent in plants: pattern formation is an integral part of development in each while hormonal influences, so striking in plants, can hardly be said to be more important in their growth than in animals; in one respect perhaps, viz. in its sensitivity to environmental influences, does the plant differ most from the animal, e.g. dormancy, geotropism, phototropism; and the seasonal 'stop-go-stop' in development is almost unique to the plant world; there is accretional growth year after year in many plants whereas there are only some doubtful examples of this in animals; nor is there anything in animals quite like the regular senescence and regrowth of annual plants. Finally, in its regenerative capacity, i.e. growth of a whole plant from a shoot, piece of stem, root, leaf or even a few cells, the plant provides valuable opportunities for investigating the phenomena of regeneration.

FURTHER READING

Black M. (1972) *Control Processes in Germination and Dormancy.* Oxford: University Press.

Gemmell A.R. (1971) *Developmental Plant Anatomy.* London: Edward Arnold (Publishers) Ltd.

Greulach V.A. & Adams J.E. (1967) *Plants. An Introduction to Modern Botany.* New York: John Wiley & Sons, Inc.

Phillips I.D.J. (1971) *The Biochemistry and Physiology of Plant Growth Hormones.* New York: McGraw-Hill Book Company.

Ray P.M. (1963) *The Living Plant.* New York: Holt, Rinehart & Winston.

Street H.E. & Opik H. (1970) *The Physiology of Flowering Plants: Their Growth and Development.* London: Edward Arnold (Publishers) Ltd.

Torrey J.G. (1967) *Development in Flowering Plants.* London: Collier-Macmillan Limited.

19 INVERTEBRATE DEVELOPMENT

Variations in the basic design for development are well illustrated in Chapter 5 where four examples, amphioxus, amphibian, bird and mammal are described. The aim was to alert the student to other intermediate and more unusual variations as well as to awaken an interest in development. All examples dealt with so far have been vertebrates except for the amphioxus, a cephalochordate which stands near the border line on the invertebrate side. If we are to examine further possible variations in early development there is a fruitful field indeed to explore among the invertebrates. Apart from the enormous number of invertebrate species (comprising nearly 95 per cent of animal species) they are also important as useful material for research into the problems of development, differentiation, regeneration etc. and deserve our full attention. With such a variety and profusion of species, it has been no easy task to achieve taxonomic order, and a full measure of agreement among the experts is still lacking. If that be the case, then a detailed exposition of their different developments is impossible in a text like this. But even a brief excursion into the field of invertebrate development can be worthwhile as long as the wood is not lost sight of amongst all the trees.

THE BASIC PATTERN OF DEVELOPMENT

The amphioxus which provided the basic developmental pattern for the vertebrate world proves an equally useful model with which to venture amongst the more primitive invertebrates even as far back as the protozoa, unicellular organisms which are close to the stem line from which plants as well as animals originated. The sequence of events in the development of amphioxus is simple (Fig. 5.2): the zygote cleaves several times to become a morula then, with the appearance of an inner fluid-filled space, the blastula is formed; after invagination of the macromeres into the blastocoel during gastrulation, an archenteron and blastopore are established; notochord, mesoderm and neural tube are derived, each in its own way, from cells on the lips of the blastopore as the embryo grows. During development of the primitive body form, the archenteron becomes the gut, and an opening which later becomes the mouth, appears at the

192

cranial end of the gut; the blastopore, which *might* have persisted as the anus, closes and a new anus near the site of the blastopore is created; the mesoderm is the source of many different tissues and organs.

The gastrula

To look amongst unicellular organisms for development analogous to that seen in the amphioxus, our 'standard' pattern, is pushing analogy a bit too far. In the first place their reproduction is usually by simple fission (division) of the organism to give two similar daughter cells and in many cases there is nothing more to report on their development. In some forms of protozoa, e.g. amoeba (Fig. 19.1), the ingestion of food may occur through any part of the

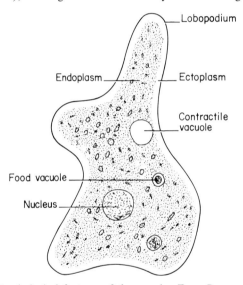

Fig. 19.1 Morphological features of the amoeba. From Barnes, R.D., *Invertebrate Zoology*, 2nd edition. W.B. Saunders Company, Philadelphia.

surface by a process resembling pinocytosis to form food vacuoles but in others e.g. Tetrahymena, Paramecium, Euplotes, an elaborate and permanent invagination of the surface membrane exists in the fully developed organism (Fig. 19.2). It acts as a kind of mouth and/or pharynx, ending blindly in the depths of the cell; from it, passage of food material occurs, again by pinocytosis, into the cell. A feature such as this suggests a comparison with the gastrula stage of the amphioxus where the archenteron and blastopore develop by invagination of the macromeres which form part of the blastula wall.

Cleavage

Amongst the multicellular organisms (Metazoa), it is easier to draw comparisons with and recognize variations from the standard pattern of development during the cleavage and later stages. For instance, they all come to acquire representatives of the three basic germ layers, ectoderm, endoderm and mesoderm, although

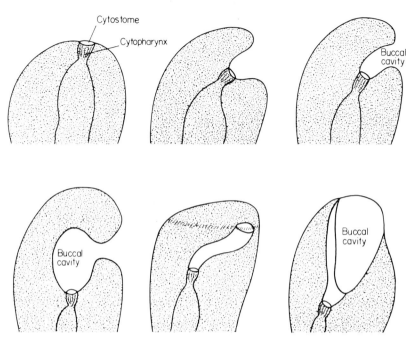

Fig. 19.2 Variations in the region of the mouth and pharynx in unicellular organisms, simulating the process of gastrulation. From Barnes, R.D., *Invertebrate Zoology*, 2nd edition. W.B. Saunders Company, Philadelphia.

the nomenclature of these may not always be the same. Only one fundamental type of cleavage has been referred to in this text so far; it is best illustrated in the amphioxus and mammal where there is a minimum of yolk in the ovum and therefore little or no interference with the process of cleavage. Not only is cell cleavage complete but the axis of cleavage is either parallel to or at right angles to the animal-vegetal pole axis; this is *radial cleavage* (Fig. 19.3) and

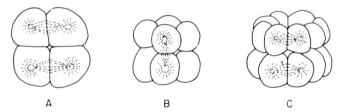

A   B   C

Fig. 19.3 The arrangement of the cells and the orientation of the cleavage spindles in radial cleavage. From Barnes, R.D. *Invertebrate Zoology*, 2nd edition. W.B. Saunders Company, Philadelphia.

variations on the theme are entirely the result of appreciable or very large quantities of yolk in the egg, e.g. in the amphibian and bird. Radial cleavage with similar variations of the pattern is also found among invertebrates, e.g. echinoderm, but in others, e.g. flatworms, annelids and molluscs a fundamentally different type of cleavage, described, perhaps mistakenly, as *spiral* can be seen; instead of the plane of cell cleavage being either parallel or at right angles to the polar axis of the embryo, it is oblique to that axis and successive generations of cleaving cells (up to a total of four as a rule) alternate between having it oblique to the right and to the left. An accurate system for designating the lineage of the dividing cells is illustrated in Fig. 19.4. While discussing

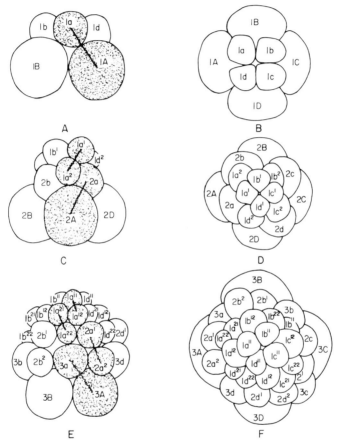

Fig. 19.4 Spiral cleavage: the arrangement and lineage of the cells in the blastula; the orientation of the cleavage spindles is shown in (A), (C) and (E) (lateral views) for comparison with (B), (D) and (F) (animal pole views). From Barnes, R.D., *Invertebrate Zoology*, 2nd edition. W.B. Saunders Company, Philadelphia.

competence and determination in relation to cell differentiation in Chapter 11, it was pointed out that, using amphibian material at the morula/blastula/gastrula stages, cells could be transferred from one part of the morula or blastula to a remote part of the same embryo without upsetting subsequent development, i.e. the transplanted cells fall into line with their new-found neighbours, being competent to differentiate along a pathway other than their

presumptive fate had indicated. Furthermore, removal of a small number of
cells from the morula/blastula/early gastrula was compensated for and no
defect was seen during subsequent development. In other words, the cells
were still indeterminate; only in the late gastrula stage was evidence of
determination in the cells obtained. In those invertebrates with radial cleavage,
the picture is the same, i.e. indeterminate cleavage, but in cases showing spiral
cleavage, each cell seems to have its fate determined at an extremely early stage;
thus the removal of one cell during cleavage is not compensated for and the loss
is obvious when the organs and tissues are differentiating. One cell in particular,
labelled 4d in the scheme shown in Fig. 19.5 and known as the *mesentoblast*

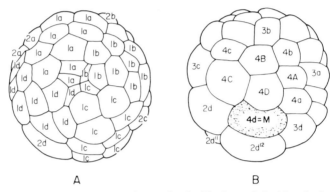

A   B

Fig. 19.5 Animal pole (A) and vegetal pole (B) views of the blastula during
spiral cleavage. The mesentoblast (4d = M) contributes all the entomesoderm.
From Barnes, R.D., *Invertebrate Zoology*, 2nd edition. W.B. Saunders,
Company, Philadelphia.

cell, is already determined as the source of a few endodermal cells and all the
mesoderm in the adult. This feature, *determinate cleavage* is regularly associated
with the spiral type of cleavage; furthermore the grouping of invertebrates into
Protostomes and *Deuterostomes* (to be explained later) also corresponds to
classifying them as spirally and radially cleaving eggs respectively. Still on the
topic of cleavage, there are of course examples among the invertebrates where
only a small amount of yolk is present in the egg and the distinction between
micromeres and macromeres may be difficult; in others, micromeres and
macromeres are perfectly obvious but there are also eggs in which, because of so
much yolk, cleavage and cell formation are confined to one side of the egg or
to the surface of the egg thus enclosing an acellular mass of yolk, e.g. in insects.

A blastula may have a large central cavity, e.g. in amphioxus, a small
excentric cavity similar to that of the amphibian or there may be no cavity
whatsoever at this stage; for instance, in Hydra, cleavage produces a solid mass
of cells of which those on the outside become ectoderm and those inside
endoderm; later a cavity representing the stomach appears in the endodermal
mass, eventually communicating with the surface through an oral opening
formed at a point where ectoderm and endoderm break down.

Gastrulation in invertebrates has numerous variations. As in vertebrates some gastrulate by invagination of the macromeres (to form endoderm) into the shell of micromeres (forming ectoderm), others by epiboly whereby micromeres envelop the macromeres but in either case a blastopore is formed; some however gastrulate by ingression, a process in which there is first a hollow blastula; the cells destined to become endoderm invade the central cavity escaping individually from the wall of the blastula to form a central mass; in this way the blastocoel may be obliterated to give a stereoblastula. Cells corresponding to, if not actually referred to as, mesoderm also develop between ectoderm and endoderm. One class (Calcarea) of the sponges (Porifera) shows an unusual feature during gastrulation; the early cleavages lead to the development of a flat disc of two apposed layers, each comprising eight cells. Those in what might be termed the upper layer are the larger cells (macromeres) although their subsequent history reveals that they become the outer layer of the organism, i.e. ectoderm. The smaller narrow cells of the lower layer are the micromeres and when the 'blastocoel cavity' develops between the two layers, each micromere acquires a flagellum projecting into the cavity. An opening out of this cavity appears amongst the macromeres and the flagellate cells in the opposite wall of the blastula push their way through this opening, thus turning the blastula inside out, i.e. *inversion.* The net result is a second blastula, again with a cavity surrounded on one side by the reconstituted layer of macromeres (ectoderm) and on the other by the flagella-bearing micromeres but the flagella are now on the outside of the cells. Subsequent gastrulation, achieved by epiboly in the Calcarea, gives first, a two-layered gastrula with macromeres on the outside, micromeres on the inside, their flagella projecting into the archenteron and, of course, a blastopore. Mesodermal cells or their representatives are later derived from the other two layers. It is sometimes confusing to come upon different names for layers of cells corresponding to the ectoderm, endoderm and mesoderm of vertebrate development but they can easily be identified by noting their ultimate position and relationships in the late gastrula.

The coelom

A great deal of attention has been given to the mesoderm and the coelomic cavities of the invertebrates in relation to the problem of the evolution of these animals but, here, we shall only note the way in which they develop, a feature which, in itself, has been important in phylogenetic theories. The origin of mesoderm varies; in sponges and coelenterates, mesodermal cells come from ectoderm while in other forms they arise from endoderm or at least from cell(s) originally sited in that layer, e.g. from the 4d cell in protostomes. The mode of development of the coelomic space also varies. As happens in the amphioxus and echinoderms, the mesoderm develops from the wall of the archenteron as a number of bulges or blisters projecting into the original blastocoel (Fig. 5.2); when they finally detach themselves from the wall of the archenteron, they form, in the amphioxus, a series of coelomic cavities each with a wall of mesoderm. Those species which have and maintain a segmented (metameric) arrangement in their bodies also retain the separate coelomic spaces (one for each segment or metamere). Where metamerism is lost, the individual spaces coalesce to give one large coelomic space. Animals which develop coelomic cavities in this way are known as *enterocoelomates*. When the mesoderm, derived from cells in the lips of the blastopore, merely lines the original blasto-coel cavity (by applying itself to ectoderm and endoderm) and thus transforming it into a coelomic cavity, the animals are known as *pseudocoelomates*. The space between the endoderm and ectoderm, in some cases, may however be completely filled with mesoderm and no coelomic cavity develops at any stage; these are the *acoelomates*. In flatworms, annelids, molluscs and in vertebrates, the coelomic cavity (or cavities in the segmented forms) develops as a space (or spaces) in the mass of mesoderm which has obliterated the original blastocoel; these are the *schizocoelomates*.

Protostomes and deuterostomes

Development from the gastrula stage onwards also shows interesting differen-ces from vertebrate development. The more primitive invertebrates e.g. Hydra, retain the original arrangement of having a mouth leading into a digestive cavity which ends blindly, even if it is often branched (Fig. 19.6). The cavity has no other opening to the outside. In flatworms, annelids, and molluscs where an archenteron is formed by epiboly or by invagination, the blastopore or part of it may remain as the mouth; if however it closes completely, an invagination of ectoderm near the original blastopore site breaks through into the archenteron to form the mouth. This arrangement i.e. the mouth deriving from blastopore or close by it, is the distinguishing characteristic feature of the *protostomes* and the origin of the name. Where an anus develops in protostomes, it appears as an ectodermal proctodeal invagination which breaks through into the archenteron either near or at a distance from the site of the blastopore. On the other hand, in *deuterostomes*, e.g. echinoderms (starfish, sea urchins), where gastrulation is usually by invagination, the mouth is formed at the anterior end of the arch-enteron (i.e. opposite the blastopore) where an ectodermal invagination, the stomodeum, meets and joins with it. The blastopore either remains open as the anus or, as in amphioxus and in the vertebrates, it closes and an anal canal is formed near it.

INSECT DEVELOPMENT

So far, we have mentioned only a few variations of the pattern of development in invertebrates but an effort has been made to emphasize the similarities between basic vertebrate and invertebrate developmental processes. One group, however, viz., the insects in the phylum Arthropoda, deserves special treatment—brief though it be—because of the vast numbers and varieties of insect species in the world and also because of their apparently marked divergence from the basic pattern of development. The insect egg is usually elongated and heavily laden with yolk and the nucleus lies in the middle of it. On fertilization, the

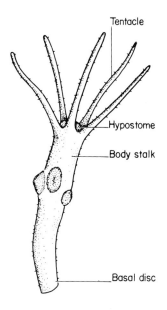

A

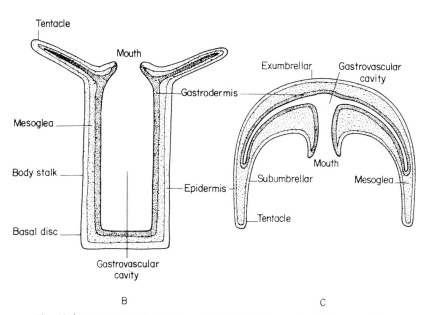

B   C

Fig. 19.6 Coelenterate body forms. (A), (B); Hydra, polypoid form, surface view and in section. (C) Medusoid body form in section. From Barnes, R.D., *Invertebrate Zoology*, 2nd edition. W.B. Saunders Company, Philadelphia.

nucleus begins dividing but very few insect eggs undergo complete cleavage; most of them show only repeated nuclear cleavage spreading throughout the central yolk mass. Most of these nuclei migrate to the peripheral region of the egg where cell membranes develop around them; a complete layer of these cells finally encloses the whole yolk mass (Fig. 19.7). Along the length of the future

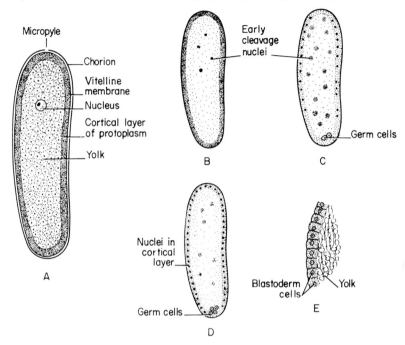

Fig. 19.7 Cleavage of the insect egg. From Ross, H.H., *A Textbook of Entomology*, 3rd edition. John Wiley & Sons, Inc., New York.

ventral surface of the embryo, these enveloping cells form a thickened *germ band* by their growth and multiplication (Fig. 19.8). Soon the band shows

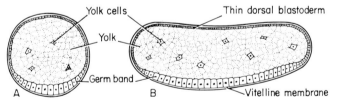

Fig. 19.8 The development of the germ band on the ventral surface of the embryo. From Ross, H.H., *A Textbook of Entomology*, 3rd edition. John Wiley & Sons, Inc., New York.

segmentation along its length while, in the midline, there is an even greater cellular proliferation resulting in an elongated mass of cells sinking under the surface to provide a layer of mesoderm lying against the germ band (Fig. 19.9). The germ band with the shell of cells enclosing the yolk forms ectoderm. At the

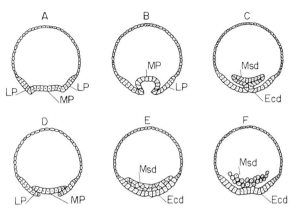

Fig. 19.9 Different methods of mesoderm formation in the insect. (A)–(C) Tubulation of middle plate (MP) of germ band with approximation of edges of lateral plates (LP). (D), (E) Middle plate of germ band sinks below level of lateral plates. (F) Mesoderm cells derived from middle plate cells. From Ross, H.H., *A Textbook of Entomology*, 3rd edition. John Wiley & Sons, Inc., New York.

anterior and posterior ends of the germ band a pocket of ectoderm invaginates to give the stomodeum and proctodeum respectively (Fig. 19.10). A small group of cells whose origin is in doubt but whose fate is to form endoderm, applies itself to the bottom of each ectodermal pocket; there, each group proliferates to form a cup-shaped, then a tube-shaped gut rudiment encircling the yolk mass inside the embryo. The gut is complete when these two rudiments meet one another (Fig. 19.10); mouth and anal canal develop by breakdown of opposing ectoderm and endoderm at each end. During the developmental stages just described, an amnion and serosa are formed over the embryo. The

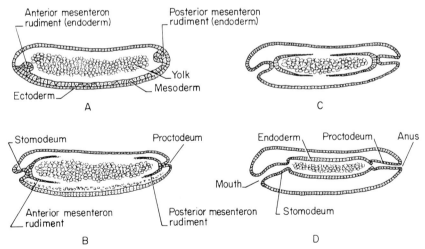

Fig. 19.10 Development of the alimentary canal in the insect. From Ross, H.H., *A Textbook of Entomology*, 3rd edition. John Wiley & Sons, Inc., New York.

mechanism is fundamentally similar to the development of the amnion and chorion over the chick embryo, i.e. a fold of extra-embryonic ectoderm and mesoderm (in the case of the insect it is the ectoderm immediately adjoining the germ band) rises up and covers the embryo (Fig. 19.11). Accompanying the

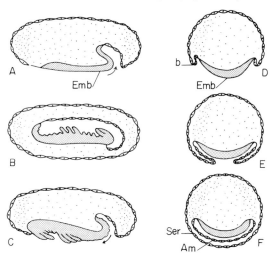

Fig. 19.11 Protection of the insect embryo. (A)–(C) The embryo slides, tail first, into the yolk and becomes covered with membranes; later the embryo reverses the process breaking through the membranes. (D)–(F) The membranes grow over the embryo which has to rupture these to escape later. From Ross, H.H., *A Textbook of Entomology*, 3rd edition. John Wiley & Sons, Inc., New York.

formation of its amnion and serosa, the embryo in some cases, performs an unusual manoeuvre; tail first, it slides into the yolk until it is completely enclosed first by amnion and, outside that, by yolk (Fig. 19.11). Later it simply retraces its steps to lie outside the yolk again. Gastrulation in the insect, i.e. the invagination of the ectoderm, origin of the mesoderm etc., as described above, is so different from our accepted basic pattern that it is often difficult to compare its stages and features with those of other invertebrates and vertebrates, e.g. invagination, blastopore etc. but to keep the basic pattern in mind nevertheless helps to understand the aberrant processes of the insect.

When segmentation has been established in the germ band the appendages associated with the segments begin to grow out, directing themselves towards the ventral surface of the embryo (Fig. 19.12). Another interesting feature in insects and other invertebrates is the site of development of the nerve cord or primitive spinal cord, i.e. on the ventral surface of the embryo where the germ band lies (see below), but its development otherwise follows that of the vertebrates. Some insects and others in the phylum show evidence of spiral cleavage but in most insects the large mass of yolk obscures that feature. Early determination of the fate of the cells in the developing insect is, however, a noticeable feature, because, during the stage of segmentation of the germ band,

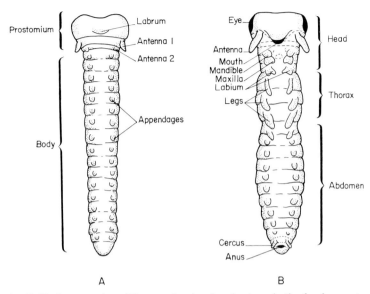

Fig. 19.12 Segmentation of the germ band and early stages in the development of appendages. From Ross, H.H., *A Textbook of Entomology*, 3rd edition. John Wiley & Sons, Inc., New York.

removal of one or more segments results in absence of the adult parts derived from those segments; there is no *compensation* or (as it is usually termed) *regulation*.

Insect development is the focus of so much study and the model for so many experiments that some of its more common features should be mentioned. After hatching the insect cannot grow smoothly and steadily like most other animals. Because of its tough inelastic skin or *cuticle* (present from the time of hatching), ordinary growth or increase in size is restricted and divided into a series of stages demarcated by *moults*, i.e. shedding of the cuticle, a process known as *ecdysis*. In the interval between each moult the stage of development of the insect is known as an *instar*. During its development, the insect has a series of moults—four, five, six or more—with first, second, third instars and so on, until it reaches the adult or *imago*. But the insect when it hatches may not be and usually is not just a miniature of the adult form; one, and usually more of its initial instars are immature and incompletely developed forms known as *larvae* which are active in feeding (Fig. 19.13). The final stage during which metamorphosis to the adult form occurs is the *pupa*, inactive, non-feeding and passed in a *cocoon* or hidden in the ground.

In other sections of this text, great emphasis was placed on the role of hormones during growth and differentiation in plants and vertebrates. The same phenomenon has been demonstrated during the developmental stages described above. Two hormones are involved: the juvenile hormone, in relatively high concentration in the organism during the early instars but diminishing towards the pupal stage, inhibits development and differentiation

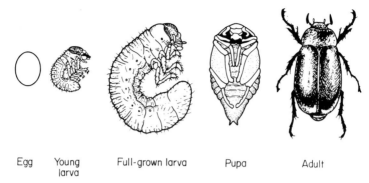

Egg   Young   Full-grown larva   Pupa   Adult
       larva

Fig. 19.13 The developmental stages of a beetle. From Ross, H.H., *A Textbook of Entomology*, 3rd edition. John Wiley & Sons, Inc., New York.

while ecdysone, which gradually increases in concentration over the same period, appears to stimulate development and metamorphosis of the insect.

ONTOGENY AND PHYLOGENY

If we are allowed to compare ontogeny and phylogeny without drawing firm conclusions or inferences from our observations, it becomes an interesting study helping us to understand more fully the complicated processes of development. For instance, the highly organized development and differentiation of the nervous system as a derivative of dorsal ectoderm will not only make sense but we may also be less disturbed at finding a nervous system e.g. a nerve cord, lying ventrally in an animal, when we examine the primitive nervous system of the early invertebrates. In Hydra the nervous system is merely a network of nerve cells found all over its body just below the epidermis from which it is derived. In more highly developed invertebrates the nervous system evolves progressively into aggregations of nerve cells to form ganglia with a tendency for these to concentrate in or near the head end and to have nerve cords emerging from them to serve the rest of the body. Later forms show such an intense aggregation of ganglia around (often encircling) the foregut that we can hardly avoid considering them as primitive brains. A series of nerve cords, often containing many nerve cells, extends from these masses along the trunk, usually one in a ventral position, one dorsally and one or more in a lateral, dorsilateral or ventrolateral position. We could regard them as forming a basic pattern of nerve cell and nerve fibre distribution which by gene mutations and natural selection has evolved into the nervous systems found in higher invertebrates and vertebrates. The single ventral nerve cord resembling a spinal cord is therefore no more peculiar and unexpected in the insect than the dorsal spinal cord in the human. We can also identify the first steps in the evolutionary development of the eye, with its pigment cells derived from the skin and retreating from the surface to form a shallow cup into which sensory nerve cells send their light sensitive nerve endings. The segmental organization of the early vertebrate 'kidneys', pronephros and mesonephros, has its origins in the metameric invertebrates, e.g. annelids; here, tubules called protonephridia, (often invaginated into blood vessels at one end) and/or metanephridia (which drain the coelomic cavity) are found for the purpose of removing water and soluble excretory products from each segment and each of them opens on the surface by its own nephridiopore. The latter feature is lost during further evolution of the excretory system; the close association of the excretory (urinary) and genital system of ducts is also an early and distinctive feature of invertebrates; both systems use the same ducts to the exterior in many cases. In vertebrate ontogeny, a relic of these primitive phylogenetic stages is the regular use of the mesonephric tubules and duct as genital ducts in the male even when the mesonephros is still functioning as a kidney. Many other examples of this nature could be quoted but it would be unwise to regard all such features found among the invertebrates as early steps in the evolution of vertebrates; safer perhaps to see them as modifications of a basic pattern of development giving a spectrum of variations from the simple to the highly complex.

FURTHER READING

Anderson D.T. (1973) *Embryology and Phylogeny in Annelids and Arthropods.* Oxford: Pergamon Press.

Barnes R.D. (1969) *Invertebrate Zoology* 2nd ed. Philadelphia & London: W.B. Saunders Company.

Chen P.S. (1971) *Biochemical Aspects of Insect Development.* Basel: S. Karger.

Counce S.J. & Waddington C.H. (1972, 1973) *Developmental Systems: Insects* Vols. 1 and 2. London & New York: Academic Press.

Horstadius S. (1973) *Experimental Embryology of Echinoderms.* Oxford: Clarendon Press.

Rockstein M. (1973) *The Physiology of Insecta* 2nd ed. Vol. 1. New York & London: Academic Press.

Ross H.H. (1965) *A Textbook of Entomology.* New York: John Wiley & Sons, Inc.

20 GROWTH OF CELLS UNDER ARTIFICIAL CONDITIONS

More commonly known as the culture or growth *in vitro* of cells, tissues, organs and embryos, these technical procedures are now widely used for research purposes in many laboratories, having already provided much valuable information on the topic of growth, development, cell differentiation, the cell cycle, inductive interactions—indeed nearly every aspect of developmental biology, and there is no doubt that further investigation and analysis of developmental problems using *in vitro* techniques will be gratifying and fruitful. Basically the techniques consist of removing the living material from its normal environment and allowing it to grow in or on a fluid or semisolid nutrient medium which is usually specific, often complex and strictly defined, accurately controlled during the experiment, kept at the correct temperature and with aseptic (sterile) conditions prevailing throughout the entire procedure.

MATERIALS AND METHODS

There are many excellent textbooks on the subject and they should be consulted for full details of the techniques but it is highly desirable before embarking on any work in this field to learn and practice the procedures in a laboratory where culture techniques are being used routinely. Here we can only give an outline of the techniques with a guide as to what is possible, what may be achieved and what may be learned. The equipment varies with the material to be grown; test tubes, watch glasses, Petri dishes and various types of flasks are usually employed to hold the cultures but special containers are required for some techniques. These and the instruments can be conveniently sterilized in an autoclave or hot air oven but it is also essential to prevent contamination throughout the experiment. Solutions (e.g. physiological saline) which are not altered or destroyed by heat are also sterilized by autoclaving as can materials like stoppers and cotton wool but many media contain blood plasma or serum which must be obtained from the animal and stored under strictly aseptic conditions. Still others, like enzymes in solution which are inactivated by heat, have to be passed through very fine filters to ensure the removal of bacteria. The commonest source of contamination during an experiment is from the air and although it is possible to carry out the procedures with the minimum of facilities for maintaining sterile working conditions, specially designed hot rooms with easily sterilized bench tops, easily cleaned floors and means of sterilizing the air are now being provided in larger establishments.

Clearly the most important and most complex element in culture work is the medium, which must be prepared to suit the cells or tissues in regard to their metabolism and nutritional requirements for survival and growth and to the physical characteristics, e.g. pH, temperature, demanded by the cells. Historically, the first medium used (in the early 1900's) was clotted lymph on which the tissue could grow; soon after, plasma clots (usually from a fowl) became a common supportive medium for this work. The exploitation of natural media such as blood serum, preferably from young animals and birds, followed rapidly and, even yet, almost all media, both semisolid and fluid, contain at least a percentage of serum because it contains and is the only source of two proteins apparently essential for maintaining the growth of cells and tissues in culture. Plasma and serum can be obtained by bleeding hens and other animals in the laboratory but it is now becoming common practice and far more convenient in most instances to buy them from commercial firms. Embryo extracts e.g. chick and bovine, are also used as additives to media because of their growth-stimulating properties.

The incorporation of even small quantities of serum or plasma in culture media demands the special care and skills necessary to prevent contamination while removing it from the animal as well as the maintenance of or ready access to stocks of all sorts of animals; the alternative is the high cost of buying the material from special suppliers. But it also leaves open the possibility, indeed the likelihood, of unknown, perhaps variable elements being introduced into the experiment without any rigid control or even full understanding. Accordingly every attempt has been made to use artificial, synthetic media prepared from physiological or balanced salt solutions containing all the ingredients e.g. amino acids, vitamins, sugars, co-enzymes etc., known to be essential for supporting cell growth and multiplication. But different cells often require their own particular metabolites, combination of metabolites and in different concentrations. Consequently a wide range of these media is available but very few cells or tissues have been successfully grown in wholly synthetic media. It is, in fact, the usual practice to add serum or embryo extract to all media in order to obtain satisfactory results. Additional protection of the culture preparation against contamination is usually provided nowadays by adding an antibiotic, e.g. penicillin, streptomycin, in small quantities. Cells and tissues find it difficult under even normal *in vivo* conditions to withstand fluctuations in the pH of their environment and growth *in vitro* also requires stability in that respect. Thus, cultures usually require an environment containing a relatively high concentration (5 per cent) of CO_2 as well as the incorporation of buffer substances, e.g. phosphate and bicarbonate in the medium. Oxygen is usually supplied too, although some cells can survive for a time without it, their energy coming from anaerobic glycolysis. The temperature of the medium carrying or containing the culture must be close to that normally enjoyed by the cells i.e. 37.5—38.5°C. for those of mammalian and avian origin but it is surprising how much fluctuation is permissible (but not advisable). A 4—5° rise to 42°C. may not damage the cells for several hours but the further the temperature rises the shorter the possible survival period. Lowering the temperature below the optimal slows growth rates but cells and tissues survive remarkably well even at a temperature of 4°C.; with the addition of certain substances e.g. glycerol, to the medium they may be stored at much lower temperatures and resume growth when thawed out. Compared with mammalian and avian tissues, those of amphibia and fish grow best at lower temperatures. A reliable incubator to maintain a constant temperature is necessary for temperatures about blood heat.

Cell culture is probably the most commonly employed technique in this field, modified and refined over recent years so that many different kinds of cells can be grown *in vitro*. The first step is to disaggregate or separate the cells from one another with the minimum of damage. The tissue is sliced and chopped into tiny fragments with a very sharp knife or scalpel. This is followed by treatment designed to digest or remove the intercellular substances, viz., by incubating the fragments for a short time in a solution containing a weak concentration, e.g. 0.5—0.05 per cent of trypsin or collagenase. Stirring or agitation of the solution accelerates disaggregation. When the fragments remaining after this procedure have settled, the solution containing the cells in suspension is centrifuged and the supernatant fluid (containing the digesting enzyme) is removed; the cells are washed in a balanced salt solution to ensure removal of all traces of the enzyme and spun down again. Clumps of cells are removed by filtration through several layers of gauze to give a suspension of single cells. A known concentration of cells, obtained by counting them in a special chamber under the microscope, is then inoculated into the prepared medium in a flat bottle, flask or test-tube for incubation at the required temperature and in an atmosphere containing the optimum concentrations of O_2 and CO_2. The cells settle on the glass surface and multiply there, spreading until a monolayer is formed. Normally a change of medium is needed approximately every three days. With many types of cells, the formation of the monolayer is followed by inhibition of growth and multiplication. It is possible however to disaggregate them by gentle washing with trypsin, to resuspend, wash, count and seed (inoculate) them again in fresh medium in another vessel—the technique known as 'passaging' the cells.

This technique allows the living cells, as they adhere to the surface of the vessel, to be examined under the phase contrast microscope and time-lapse cine-photography studies can also be carried out to demonstrate the cell movements which are greatly accelerated when the film is projected at the normal running speed. More detailed studies of cell structure and relationships can be achieved by fixing the cells and examining the stained preparations particularly on small coverslips placed in the container before inoculating the medium and removed after the cells have adhered to and grown over them. The cells may also be examined by the EM, often after removing them mechanically, i.e. by gentle scraping from the glass surface after fixation; the material obtained in this way is spun down to form a pellet of cells which can be processed and sectioned on the ultra-microtome. One disadvantage of this procedure is the possible cell damage caused by their removal mechanically from the surface of the vessel. Removal by trypsinization, however, cannot ensure that no harm has come to them in relation to their fine structure. This basic technique of cell culture offers innumerable opportunities for studying and investigating various phenomena, e.g. cell movements, cell differentiation and intercellular reactions, cell aggregation, the effects of different drugs, various metabolites and hormones added to the medium, the synthesizing activities of the cell as estimated by the incorporation into the cell of specific radioactive tracer substances added to the medium, etc. It must be emphasized at this point that the results obtained using *in vitro* techniques do not necessarily apply to *in vivo* conditions.

There are many modifications of the technique; it is possible, for instance, to inoculate the medium with tiny chopped fragments of tissue e.g. liver, muscle; when these adhere to the surface of the vessel, the cells, epithelial, fibroblast and muscle, grow out from the fragment and can be studied *in situ* or removed by trypsinization for further culture as single cells. Agar slope cultures in test-tubes are also used; this semisolid medium comprises a solution of agar and culture medium, warmed and poured into a test-tube tilted to give a long surface when the agar cools in that position. A small quantity of medium is run over the surface and cells, usually from an other culture, are transferred by means of a platinum wire loop.

Some kinds of cells can be grown and maintained *in suspension* in the medium as opposed to allowing them to settle and grow as a monolayer. It is a useful technique if large quantities of cells are required or if biochemical estimations have to be carried out on cells. The media employeu are essentially the same as for other techniques with substances like methylcellulose incorporated to maintain the cells in suspension. The more important modification of the basic culture technique however is to keep the medium moving by gently shaking or rotating the flask or drum throughout incubation or by using a magnetic stirrer lying in the bottom of the flask and activated by a small electric motor incorporated in the base which supports the flask.

There are two main types of cells or *cell lines* in culture work, judged according to whether or not they can be passaged or subcultured an infinite number of times. The technique outlined above for growing cells derived directly from normal tissues gives a *primary culture* which, when subcultured, gives a *primary cell line*; in this case the cells retain their normal characteristics. A good deal of care and rich media are needed to maintain their growth in repeated passages but in most instances cells, after a certain number of passages, eventually stop multiplying and die in spite of every effort to prevent them doing so. Some kinds of cells obtained from normal tissues, appearing and behaving initially as primary cell lines undergo *transformation* or *cell alteration* to become *established cell lines*; the cells change in their appearance, behaviour and genetic constitution (i.e. becoming polyploid or aneuploid), show rapid rates of multiplication, can be passaged indefinitely and can therefore no longer be considered 'normal'. Unfortunately little is known about the mechanism whereby primary cell lines become established lines but viruses, carcinogenic substances and X-rays are all credited with being able to cause cell alteration (probably an alteration of the genome) in a primary line. Many established cell lines, each specially named and/or numbered e.g. HeLa, CCL2 and BHK 21, CCL10, have been produced and are in common use in laboratories throughout the world as experimental material. Many established lines were derived from normal (often human) tissue while others were obtained from cancerous growths, e.g. the HeLa cell line came originally from a carcinoma of the human uterine cervix. Cell culture is already a popular research technique in nearly every field of biological research and its potential is far from being exhausted. Yet we have still a great deal to learn about the 'normal' behaviour of cells in culture e.g. their metabolic characteristics, degrees of dedifferentiation and differentiation, intercellular reactions and cell aggregation phenomena.

Tissue culture

The first method of growing living material *in vitro* was by *tissue culture* and the techniques are still in use although not so widely as the more recently introduced techniques of cell culture. Essentially the method consists of placing a tiny fragment of tissue on a plasma clot, to which it adheres, and allowing it to grow in a small moist chamber. For instance, a 'well' or depression slide (one with a hollow depression in it) may be used as follows: a drop of plasma is allowed to clot on a coverslip, the tissue fragment is placed on the clot and moistened with embryo extract. The coverslip is then inverted over the hollow of the slide and sealed in position with sterile petroleum jelly. An advantage is its simplicity for short term studies (with the microscope) of the cells growing out from the tissue fragment. Several modifications are in current use, e.g. the plasma clot carrying the fragment(s) may be adhering to the wall of a test-tube, on a coverslip inside a test-tube or on the floor of specially made flasks (e.g. Carrel type); a drop of culture medium may replace the plasma clot on the coverslip which is inverted over the hollow on a depression slide i.e. the 'hanging-drop' method. As mentioned when the cell culture technique was being described, it is also quite easy to inoculate a liquid growth medium with tissue fragments and allow them to settle on the surface of the vessel; the cells grow out as in the simpler hanging drop or plasma clot slide cultures.

Organ culture is aimed at obtaining normal growth or even cell survival throughout a whole fragment of an organ under *in vitro* conditions without migration of cells from the fragment. Originally embryonic parts were the subject of study, e.g. for the growth and differentiation of an embryonic limb bone; plasma clots were used to carry the embryonic material but repeated transfers to new plasma clots were obviously a hazard and now the culture material is usually supported on a raft of lens paper or synthetic fabric. This technique allows the use of a fluid nutrient medium on which the raft can float and not be submerged in or covered by the medium; it is sufficient to have the preparation in a moist atmosphere within the culture chamber. Fine metal grids in the form of a platform keeping the cultured tissue just moistened by the medium is a further modification. Fragments of adult organs are more difficult to grow *in vitro*; they usually require a higher concentration of oxygen (up to 95 per cent) in the surrounding atmosphere and serum is usually omitted from the growth medium but otherwise the technique is similar.

Embryo culture is as fascinating and in many cases as challenging as any other type of *in vitro* work; the aims are usually to study closely, and if necessary continuously, the normal development of the whole embryo and also to study development after surgical interference or treatment with drugs and chemical substances likely to upset the normal process. Amphibian and fish embryos present little difficulty apart from providing conditions in the laboratory similar to those in which development normally occurs, attention being paid to temperature, lighting, aeration and particularly to the composition of the water, e.g. ordinary tap water is not always satisfactory because of its chlorine or metallic ion content.

The culture of avian and mammalian embryos needs far more care. It is possible to open a hen's egg at different stages early in the incubation period and, provided the opening is protected from infection and drying, the development of the embryo may be followed for several days but close study of the embryo is not feasible. The first attempts to grow the chick embryo outside the egg followed the techniques employed for cell and tissue culture; the embryonic disc was removed from the egg at 24–40 hours incubation and spread on a plasma clot. Later, a semisolid medium consisting of agar and egg albumen or of agar and whole egg was introduced to support the growth of the embryo. In this way the development of the embryo continues normally for a further 24–36 hours in a moist sterile atmosphere and can also be recorded by time-lapse cinephotography. The results of interfering with embryonic growth can also be closely studied under such controlled conditions and compared with *in vivo* work. Equally good and perhaps better results can be obtained by stretching the vitelline membrane (carrying the 24-hour embryonic disc) over a glass ring and supporting the preparation on egg albumen alone. Unfortunately, techniques for obtaining normal growth, *in vitro*, of the chick embryo beyond the third or fourth day have not yet been devised.

In spite of the obvious difficulties encountered in the retrieving and handling of mammalian embryos, there has been remarkable success. With mouse, rat and rabbit material, it is now possible to remove and culture embryos from the single cell zygote up to the blastocyst stage. The length of time which they may be grown *in vitro* depends on their age when removed from the mother, longer in the case of the early stages because there is great difficulty in taking them beyond the blastocyst stage. Much of the work on these early pre-implantation embryos has been done in association with experimental transplantation of the embryos from the uterus of one animal to that of another, i.e. to a uterine foster mother of the same species or even, for temporary storage, to a mother of a smaller species and then back to a foster mother of the same species e.g. sheep—rabbit—sheep. Special techniques have been introduced and are now being elaborated to culture post-implantation rat embryos enclosed in their chorionic sacs from the early somite to the 40-somite stage.

Mention should be made here of the techniques using the hen's egg for growing cells and tissues (usually embryonic) outside their normal environment. A popular and easy technique, although it requires strict aseptic measures is chorio-allantoic grafting. A window is cut in the shell and shell membrane of an egg which has been incubated for ten days and the fragment of tissue or cell suspension is placed on the surface of the exposed vascular chorioallantois; after the window has been closed (by shell or sterile material) and sealed, the implanted material can grow to an advanced stage during the ensuing seven days. A more difficult technique involves the implantation of tiny tissue fragments into the *intra-embryonic coelom* of the $3\frac{1}{2}$–4 day chick embryo; the period of growth for the graft is longer than with the chorioallantoic technique and the tissue need not be retrieved until the chicken hatches. Both these techniques have limited applications compared with those mentioned earlier but they do provide easy demonstration of differentiation in embryonic tissues.

Culture of plant material is also firmly established particularly for the analysis of growth processes and their metabolic requirements. The nutrient media vary greatly according to the species, consist usually of a selection of salts as well as sugars and amino acids and may be in a liquid or semisolid form (i.e. with the inclusion of 0.5% agar); for many purposes however, the addition of endospermic material such as coconut milk is essential. Sterility and aseptic conditions are also important in plant culture work but no special 'gassing' of the medium is required and the material grows readily at room temperature.

The aseptic removal of plant tissues from storage roots, e.g. carrot, from stems or branches and from seeds and fruits needs care; the outer exposed surfaces must be carefully cleaned with antiseptic or disenfectant before the deeper tissues are approached by cutting or breaking. The tissue for culture is removed aseptically and placed in or on the culture medium. No attempt is made to disaggregate the cells at this stage. In culture, proliferation of the cells in the explanted tissue gives rise to irregular, rather friable masses known as callus. It is possible to have shoot and root meristems form inside a callus but callus can be broken up by shaking or rotating the flask to free the cells individually or in groups and these will continue to proliferate in suspension media. With carrot or tobacco plant tissue as the initial explant and with coconut milk incorporated in the medium, the individual cells from callus may form embryos which can be reared to give normal adult plants. This is a greater achievement, i.e. the development of embryos from somatic cells and their progression to a whole plant, than has been possible with animal tissue, reflecting, perhaps, the greater regenerative and dedifferentiative capacities of plants in general. It is interesting to note that the endospermic material is necessary for these procedures and may be compared with the need for embryo extract and/or fetal serum in so many animal tissue culture media.

Plant embryos can be removed from seeds and grown successfully *in vitro* to the adult stage as well; the embryo must be beyond the 'globular' stage of its development. The nutrient requirements are more exacting for embryo cultures than for somatic cells. Culture of organs such as roots, shoot apical meristems and leaf primordia is also possible in sterile media.

FURTHER READING

Austin C.R. (1973) *The Mammalian Fetus in vitro.* London: Chapman & Hall.

New D.A.T. (1966) *The Culture of Vertebrate Embryos.* London: Logos Press Limited.

Paul J. (1970) *Cell and Tissue Culture* 4th ed. Edinburgh & London: E. & S. Livinstone.

Thomas J.A. (1970) *Organ Culture.* New York & London: Academic Press.

Willmer E.N. (1966) *Cells and Tissues in Culture* Vol. 3. New York & London: Academic Press.

21 PROSPECTS

Those who have read this far in the text will now realize the breadth and ramifications of developmental biology. As for the basic principles of growth and development—well, we have still to define and elucidate many of them. Just how long it will take to solve those crucial problems will depend on the ability and enthusiasm of the student and researcher. There is however the consolation that developmental biology is already an active field of research and, with so much scientific 'know how', both factual and technical at our disposal, the future seems reasonably assured.

And for what purpose, it may be asked, should one strive to discover all the minutiae of growth and development when the process can and does continue so effectively and autonomously in spite of us? A passing acquaintance with developmental biology may be all that is needed to have a better appreciation of the slow, sometimes imperceptible, sometimes dramatic changes in all living material from the microbial populations to plants and trees, the insects, the domestic animals and in ourselves. But if it is important, in any of these spheres, to be able to influence the life cycle, be it for pleasure or profit, then a better understanding of development would definitely be an advantage. Should an interest in developmental biology for its own sake be adopted as a scientific study with some expectations of achievement, then the prospects are exciting because there are many mountains to be climbed.

With the congenital abnormality as a *fait accompli* when first seen, the only alternatives are either to prevent its recurrence or to provide treatment, correcting it efficiently, preferably by applying the principles of normal tissue growth. In the former context, the problems of the teratologist were outlined in the text and are largely a matter of defining precisely the metabolism of drugs and toxins in association with the variations in the delicate, though unknown, reactions occurring in cell differentiation in different tissues. Treatment of defects is even more of a challenge; these often arise from failure of growth in some part of the body and the ideal therapy would be to stimulate and accelerate growth and differentiation in the tissues until it reaches its 'normal' size for its age and can continue in step thereafter.

Even more of an achievement would be the ability to 'grow' or 'cultivate' fully differentiated tissues *in vitro* from a small number of cells belonging to the future host for the restoration of a congenital defect. If this ever becomes possible, it should not be much more difficult to cultivate tissues and organs from the host for the host to replace damaged and diseased organs like the heart, liver or kidneys particularly to avoid the immunological responses of normal transplantation procedures. We are far from seeing these dreams being realized but they should not be considered for ever impossible.

Commonly held ideas on the ultimate achievement of developmental biology concern the 'test-tube' baby, an animal reared artificially from fertilization to independent existence. Work is already being done in this direction using laboratory animals and requiring a great deal more ingenious equipment than merely a test-tube. The immediate prospects of achieving a period of *in vitro* growth and development corresponding to the whole of gestation are still poor. Fostering, i.e. the transference to, and growth of ova or early cleavage stages of the mammalian ovum, in another uterus is already a frequent experimental and veterinary procedure and probably it has already been achieved in humans.

But these are the Everests in the far distance; there are still many foothills to be conquered. To be able to control and manipulate growth and differentiation for the benefit of the individual; to give the optimum rate of growth for longevity and to stave off ageing; to learn the mysteries of the developing brain and explain the variations in behaviour which are accepted as normal and those which are not; to contribute effectively to the problem of cancer where the controlled normal growth of cells in a tissue has been lost. To be able to grow more food (both animal and plant) more economically, would perhaps be an even more altruistic exercise, particularly in respect of the sea and other waters in which growth and development still cannot be controlled to the benefit of man.

No one can forecast whether or when any of these dreams will ever be realized but nothing will ever be accomplished without idealistic aims in study and in research.

INDEX